Cranes

A Natural History of a Bird in Crisis

Cranes

A Natural History of a Bird in Crisis

Janice M. Hughes

FIREFLY BOOKS

A FIREFLY BOOK

Published by Firefly Books Ltd. 2008

First printing

Publisher Cataloging-in-Publication Data (U.S.)
Hughes, Janice.
    Cranes: a natural history of a bird in crisis / Janice Hughes.
[256] p. : photos. (chiefly col.) , maps ; cm.
Includes bibliographical references and index.
Summary: An examination of the evolution, biology and environment of all 15 species of crane. The whooping crane is featured
as an example of a bird battling its way back from the brink of
extinction.
ISBN-13: 978-1-55407-343-6
ISBN-10: 1-55407-343-X
1. Cranes (Birds). I. Title.
598.31 dc22   QL696.H8447   2008

Library and Archives Canada Cataloguing in Publication
Hughes, Janice Maryan, 1958-
    Cranes : a natural history of a bird in crisis / Janice Hughes.
Includes bibliographical references.
ISBN-13: 978-1-55407-343-6
ISBN-10: 1-55407-343-X
    1. Cranes (Birds). 2. Whooping crane. 3. Endangered species.
I. Title.
QL696.G84H84 2008   598.3'2   C2008-900805-7

Published in the United States by
Firefly Books (U.S.) Inc.
P.O. Box 1338, Ellicott Station
Buffalo, New York 14205

Published in Canada by
Firefly Books Ltd.
66 Leek Crescent
Richmond Hill, Ontario L4B 1H1

Produced for Firefly Books Ltd. by:
INTERNATIONAL BOOK PRODUCTIONS INC., Toronto
Cover and interior design by: Dietmar Kokemohr
Edited by: Barbara Hopkinson

Printed in China

*The publisher gratefully acknowledges the financial support for our
publishing program by the Government of Canada through the Book
Publishing Industry Development Program.*

For my mother, who showed me the way

and then set me free

As do all crane species,
red-crowned cranes
renew pair bonds with
their life partner through
elaborate courtship
display and unison calling.

Table of Contents

Introduction

*I*f science fiction were reality, humans would be the consummate terraformers, modifying other worlds to support their own existence. Yet in our more tangible 21st-century tragedy, we humans have not stretched out our hands to the stars to sustain the burgeoning masses, we have merely fed unsustainably upon our home resources. The victims of our actions are not alien races, but the two million plant and animal species with which we share this small blue planet.

Since human survival became a dominant ecological process on earth a hundred millennia ago, other species have inevitably borne the burden of our extravagances. The first evidence of this occurred during the cyclical glaciations of the Pleistocene epoch, when over 80 percent of the world's megafauna — animals greater than 100 pounds (44 kg) — disappeared from the face of the earth. Extraordinary creatures such as mammoths, giant ground sloths and saber-tooth cats fell systematically, coincident not with climatic change but with the arrival of humans. These extinctions were followed by the loss of island species — moas, elephant birds, giant lemurs — that had evolved in secluded paradise, blissfully ignorant of a predator's needs. Anything that was large enough to feed a family or slow enough to kill with little effort was eaten to extinction.

Soon, however, humans would invent the means to take down those animals once too swift or too wily to grace the dinner table. We also learned to adapt the land for our own uses, swallowing up vast tracts of diverse habitat and converting it to a veritable biological wasteland. Known as the "sixth wave," we now bear witness to a mass extinction crisis that rivals the severity of five great extinction events that happened deep in the geological past, during the Ordovician, Devonian, Permian, Triassic and Cretaceous periods. Biologists have suggested that the rapid loss of species experienced in the past five centuries is at least 1,000 times higher than expected background extinction rates occurring gradually through species turnover. Unlike the first five waves, which can be attributed to natural processes such as volcanism, asteroid impact and plate tectonics, the current extinction phenomenon must be blamed almost entirely on humans. At our hands, the world's biota is quickly succumbing to habitat loss, overexploitation, pollution, invasion of exotic species and global climate change. The World Conservation Union currently lists 16,119 species at risk of extinction in the near future. Among these threatened plants and animals are one in three amphibian species, one in four mammals, one in eight bird species and one in four coniferous trees. Standing tall amid these desperate numbers are nine species of cranes.

Sarus crane populations have been declining steadily for many decades, and are now considered vulnerable to extinction due to wetland habitat loss, agricultural and industrial development, pollution and warfare.

Cranes exhibit very slow population growth because, typically, only one of two chicks hatched by each breeding female in a given year will survive.

Cranes are among the most severely threatened of all bird families. Virtually all 15 species have suffered significant reductions in their geographic distributions and population numbers. Whooping, red-crowned and Siberian cranes are currently listed as endangered. Six other species are considered vulnerable to extinction, and another is near threatened. The reasons cranes have suffered more than other avian groups are complex. They have been exploited for trade and commerce, and they have conflicted with humans for agricultural resources. Many species are reliant on the world's endangered wetlands for food and nest sites. Moreover, most cranes are migratory and, thus, have requirements for survival that often span thousands of miles, from their northern breeding territories to southern wintering grounds, and include each night's stopover en route. Their world understands no political boundaries yet they are subject to the whims of local, regional and national governments in times of both decline and recovery. Finally, cranes are constrained by their own biology, for they have evolved in tune with a life history strategy that favors long life and slow population growth, one that is fully at odds with the anthropogenic crises that they confront.

In our comfortable North American existence, we have come to believe that all unpleasant things happen somewhere else. Our small planet seems adequately large when laying blame for what we see on the evening news; the greatest affronts to our earth — overpopulation, exploitation, environmental disaster and mass extinction — must certainly be the fault of another hemisphere. However, not all threatened species can be so easily swept under the rug, particularly not one so magnificent as the whooping crane. This most endangered crane was not made so out of the desperate need of abject poverty, or even the ignorance of its biological merit. The North

American people created one of the world's most endangered species and, for a time as it declined rapidly into oblivion, they knowingly hastened the process. The plight of the whooping crane is an epic tale, not bedecked in pomp and circumstance, but laced heavily with the injuries of cruelty and indifference. Yet all good stories have heroes and villains, and whooping cranes have known both. This book will unravel the tale of their near extinction and their slow climb back from the brink as it reveals those humans that played a hand in the journey; for this, too, is their story.

The lovely demoiselle crane occurs in six primary populations that traverse 47 countries as they migrate between their breeding and wintering homes.

Chapter 1
The Natural History of Cranes

When we hear his call, we hear no mere bird.
We hear the trumpet in the orchestra of evolution.
— Aldo Leopold —

Cranes are ancient in origin. Long in neck, leg and wing, they are large, imposing wading birds that are among the tallest of the world's bird species. Cranes are found on all continents, except South America and Antarctica, and those species that breed in North American and Eurasian northern realms undertake long migrations each spring and fall. They are typically associated with open wetland and grassland habitats, where their bright plumage, graceful proportions and convivial nature are displayed to excellence in elaborate dancing and duet calling. They choose lifelong mates and are devoted parents that raise their young with both tenderness and determination.

Field guides frequently compare cranes to other wading birds — they are not day-herons, which fly with retracted necks; they are not egrets, which are much smaller in size; they are not storks, which have shorter legs and longer bills. Yet, to focus on what they are not is an injustice to all that cranes embody, because what they truly are is like nothing else on earth.

Mythology, Folklore and Cultural Importance

Cranes are at the heart of legend in nearly all human cultures graced by their presence. Perhaps this is because they somehow personify what we, as humans, truly aspire to be. Cranes are bestowed with beauty and grace, and through our eyes we see their dance and song as an expression of life's joy, enduring devotion to each other, and an anticipation of love and life passed with one's soul mate. Moreover, cranes are fiercely protective of family, yet gentle and nurturing to their young, with an innate ability to teach them well and then set them free without remorse or sentiment. Although we understand the perils that birds overcome on their semiannual journey to other lands, we do not wish to explain migration merely as behavior molded by the hands of evolution, because to us, an earthbound species, it is almost an unearthly reflection of strength and endurance that validates the will to survive, and an unabashed display of the true gift of freedom. For these reasons, we have endowed cranes with supernatural aspects.

The elegant beauty of cranes shines in this woodcut depiction of red-crowned cranes, or tanchos, printed by renowned Japanese artist Hiroshige in 1857.

Through place and time, we have both loved and revered cranes. In North America, many First Nations peoples chose cranes as totems to represent tribal clans or kinship groups. The Anasazi, Hopi and Zuni Pueblo peoples of the southwestern United States each harbor a crane clan, and the images of flying and standing cranes, and their tracks, have appeared in petroglyphs and pictographs (rock carvings and paintings) and ceramics since the year AD 900. Furthermore, the Ah-ji-awak is one of seven original clans of the Ojibwe people of eastern North America's Great Lakes region. Their legends describe how the Great Spirit, Gichi-Manido, sent Crane to earth to make its life there. When the great bird saw the beauty of the earth and the generous bounty that it offered, Crane sent forth a loud solitary call that summoned the bear, catfish, loon, moose and marten clans to join it in community. The Ojibwe clans gave the people a system of government and division of roles and labor in society; people of the crane clan have been traditionally bestowed the honor of tribal leaders.

Ancient Greeks demonstrated their reverence for cranes in an unquestionably academic fashion. Legends maintain that several new Greek letters were invented by the mythical hero Palamedes as he watched cranes in flight: the V-formation was depicted by the letter *lambda*, and *alpha* and *upsilon* described the shapes that cranes make when they are changing positions within the flock. Not to be stingy, Ancient Roman author Martial and fifth-century statesman Flavius Cassiodorus both declared that the entire Greek alphabet was obtained from the flight of cranes by the god Mercury. Cranes have also influenced language elsewhere in Europe. For example, the Latin word *congruere* (= congruence; *grus* = crane), meaning an agreement, was derived from the "unified and collective" behavior said to be exhibited by cranes. A growing interest in genealogy, the study of lineages, occurred in England after the Norman Conquest in 1066, from which was spawned a late Middle English word to describe the branching diagram representative of a family tree: *pied de grue* (= pedigree), meaning crane's foot.

Europeans through the ages accorded cranes badges of valor and propriety. The Romans were well familiar with the movements of great crane flocks, and used them to mark the changing seasons. They supposed that cranes assembled before departure according to rank and order and flew in formation reflective of this understanding, although, unlike tyrants or kings, flock leaders would voluntarily give up command by changing places with other birds, thus symbolizing equality, democracy and freedom among fellows.

One cannot neglect our historical appreciation for the beauty of cranes. They were commonly depicted in pastoral scenes on tapestries dating from the Middle Ages. Five millennia ago images reflecting their dance adorned rocks in western Sweden, where people still celebrate dancing cranes as harbingers of spring. In ancient Greece, the circular movements of the dance also symbolized the change of seasons as the passing of time. The arrival of cranes in spring represented the sungod's resurgence, and people danced the *geranos*, or crane dance, to acknowledge the circle of life that embraces both fertility and death. Crane dances continue to be performed at funerals by the Oskits peoples of Siberia.

Ancient Egyptians also carved and painted scenes of cranes on the stone walls of their tombs and temples, although these were not wild birds. The artists illustrated how Eurasian and demoiselle cranes were captured with large, hexagonal nets during migration and raised in captivity for food or, perhaps, as pets to rid their houses of vermin, as do other African peoples today. Carvings at the temple of Deir-el-Bahri depicted cranes walking between slaves with their bills tied down to their necks to prevent them from flying. Their image in burial chambers guaranteed the deceased an abundant food source in the afterlife. Yet the Egyptians also worshiped cranes on a more spiritual plane. The *benu* was a mythical sunbird deity — depicted as a crane in early Egyptian history — that symbolized the rise of life as it reformed itself anew each morning at daybreak. The bird, originally associated with the sun god Ra, later became a manifestation of the death and resurrection of Osiris, the god of the dead. Some historians believe that pervasive legends of the phoenix, the mythical bird that sets itself aflame and is reborn from its ashes, are originally derived from the Egyptian *benu*.

In other African cultures, cranes are revered for the beauty of their appearance, voice and dance. The Bambara people of Mali believe that the splendor of crowned cranes confirms that they possess self-knowledge of their gifts, and that their wisdom should not be neglected. They deem that cranes were present at the birth of speech, as the first words of their language were being created, and that mankind learned to speak by copying the cranes' voice. The traditional Kenyan story of Arap Sang and the cranes recalls how crowned cranes were granted their lovely golden crowns after their gestures of kindness saved the life of demigod and chief Arap Sang. But the cranes taught Arap Sang a lesson in the great responsibility of gift giving, for his original presents were fashioned of solid gold, which caused the birds to be hunted by people of surrounding villages. Arap Sang recognized his misjudgment in time and replaced the crowns with halos of golden feathers. Crowned cranes still maintain iconic importance in Africa as the national birds of Nigeria and Uganda.

In Australia, the brolga crane is the subject of legend, art and ceremony among Aboriginal peoples. Their culture is rich in stories describing the mythical transformation of this charming crane from its namesake, Brolga the dancing girl. In the dreamtime, Brolga was renowned for both the beauty of her merry nature and her love of dance. Many, including the wicked magician Nonega, sought her hand in marriage. When she refused to marry him, he vowed to seek revenge. One day when Brolga was dancing in an open plain, she was engulfed in a whirling cloud of dust. When it had passed, the village people saw a beautiful gray bird dancing as Brolga

Brolga cranes are the namesake of a charming young woman who danced during the dreamtime of Aboriginal Australian legend.

had danced. The crane bearing Brolga's name still dances on the grassy plains of tropical Australia.

Although cranes are honored throughout the world, they may have most deeply touched the hearts of Asian peoples. Nine crane species grace their landscape, and it is among the cultures of China and Japan that they are, perhaps, paid the greatest homage.

The cranes of Chinese legend stretch beyond the physical realm and are often depicted as creatures that span the distance between bird and human, or mortality and immortality. In 1900, a Taoist monk was restoring frescos on stone walls in the Caves of the Thousand Buddhas in Tun-huang, northwest China. Behind these walls, he discovered a secret chamber enclosed with brick to prevent its contents from being pillaged by invaders. Concealed in this chamber were hundreds of manuscripts dating from the fourth to tenth centuries AD, including a best-loved fable, "The Crane Maiden." This story is one of many transformational tales that teach the virtue of compassion, when a kindly peasant helps a wild crane. In gratitude, the crane becomes a beautiful young maiden who stays with her rescuer as his wife. The woman, however, longs for her former life as a crane. Although she is filled with remorse for abandoning her human child, she dons her heavenly robe and returns to her rightful place in the sky.

Chinese legend also describes cranes as messengers of gods who carried the souls of the dead to paradise. Sages, ancient men of wisdom, could fly to heaven if they rode on the backs of cranes, which were known to travel long distances at great heights, but if a sage could transform himself into a crane, then he could gain true immortality. Thus, the crane's image came to symbolize spirituality, wisdom, nobility and superiority. It was embroidered on the robes of high officials in imperial China, and statues of cranes flanked the emperor's throne in Beijing's Forbidden City. The birds' grace and dignity, as befitting the wise, were associated with scholarship. Beautiful handmade lutes were highly prized by their owners. Many songs were inspired by the flight and calls of cranes, and it was considered most proper for people to name their lutes with crane motifs, including *He-you* (Friends of the Crane) and *Ling-xiao-he-li* (a Crane Crying in the High Air). Perhaps it is most appropriate that the earliest complete, playable multinote musical instruments were recently discovered by archeologists at Jiahu, in Henan province. These small flutes had been carved from the hollow ulnas (wing bones) of red-crowned cranes, and date from nearly 9,000 years ago.

The magnificent tancho, or red-crowned crane, may be the most auspicious symbol in Japanese art and culture through the ages. As a universal icon of goodwill, this crane has long personified fine health and longevity, fidelity, good fortune, courage and hope for peace. As in many other cultures, the wish to aspire to these qualities spawned innumerable legends throughout Japanese history. Japanese fables frequently portray cranes as heavenly or magical beings that take human form to bestow their grace on a worthy person, often to teach two lessons: the worth of kindness and the bitterness of betrayal.

Both "The Crane Daughter" and "The Nobleman and His Sword" are stories of cranes who become women who devote their lives to another, provided a promise is sacredly kept. Unfortunately, their happiness together is short-lived; as the promise is broken, the crane forsakes her human form and returns, in despair, to her own kind. Thus, another quality embodied by the tancho is woven into Japanese cultural myths: fidelity. It has long been known that cranes mate for life and, as such, they have been revered as symbols of faithfulness and loyalty in marriage since the ninth century. This reflection is not lost on today's culture; crane motifs are still embroidered on wedding kimonos in Japan, and the tancho is a popular theme for wedding gifts and festivities.

Japanese myths depict the great longevity of cranes as perhaps measuring a thousand years or more. This notion may have arisen from the belief that cranes somehow traverse the boundaries between earth-bound mortals and heavenly creatures; to live a millennium would appear almost immortal. This conviction may have gained additional fodder through 12th-century science with, perhaps, the first bird-banding project. Legends tell how Yorimoto, Japan's first shogun, attached labels to cranes' legs. He requested that anyone finding a banded crane should record their location on the label and then release the bird; thus, he garnered valuable information on the seasonal movements of cranes. What is more, people claimed to have seen many of Yorimoto's labeled cranes alive many centuries after the bander himself had died: a testament to crane longevity, no doubt.

Red-crowned cranes have granted the promise of courage to their people down the ages through both war and peace. An ancient Japanese creation story describes a legendary hero whose deeds were so honored that upon his death his soul became a crane and took flight. Historically, hundreds of cranes have been released as thanksgiving following a victorious battle. A paper prayer was tied to the leg of each bird so that the bequests of the fallen could be carried to heaven.

As such, cranes in multitude became a plea for hope and a wish for good fortune. Perhaps the first use of the "thousand crane" motif in Japanese art was an enormous scroll painted by 17th-century artist Tawaraya Sotatsu. Measuring almost 50 feet (15 m) across, the scroll illustrates a large host of delicate silver cranes dancing along a golden shore or gliding though fair clouds. This inspiring work ensured that the thousand-crane theme would live beyond the parchment edges. It was eventually adopted into origami, the traditional Japanese art of paper folding, and a legend was born promising good health and long life to anyone who could fold 1,000 paper cranes.

Perhaps there is no story more poignant than that of a young Japanese girl and her thousand cranes. Sadako Sasaki was born in 1943, just two years before the atomic bomb was dropped on Hiroshima. Indeed, this was the event that sealed the fate of some 200,000 citizens of that city, including Sadako herself. As she grew up, Sadako appeared healthy and strong, but at age 11 she was diagnosed with leukemia caused by exposure to radiation. Although Sadako could see the misery surrounding her, she never gave up hope; she vowed to fold 1,000 paper cranes so her wish for good health and eternal peace would be granted. Visitors to her hospital bed brought her colored paper to fold and Sadako's brother hung the finished cranes from the ceiling of her room. Unfortunately, Sadako did not achieve her goal. She died on October 25, 1955, at age 12 after completing 644 paper cranes. In her honor, her school friends folded the remaining 356 cranes so that Sadako could be buried with 1,000 cranes at her side. Like the tancho itself, Sadako's wish has lived beyond her mortal life; her story has become a symbol for the hope of world peace. After

Sadako's death, the children of Japan and nine other countries raised money to build a memorial to those who had perished by the atomic bomb. In 1958, the Children's Peace Monument — a statue of Sadako holding a golden crane — was unveiled in Hiroshima Peace Park. Each year, 10 million paper cranes are sent by children around the world to be offered before the monument in hope that Sadako's plea for global peace will be answered.

I will write "peace" on your wings and you will fly all over the world.
— Sadako Sasaki —

Classification and Evolution

All 15 crane species are classified in the avian family Gruidae, whose scientific name may have been derived from the Greek word *geranos*, which means "crane." According to ancient Greek myths, a proud and beautiful queen named Gerania convinced her people to neglect all other gods so that they might worship only her. As punishment for her excessive vanity and self-indulgence, the high goddess Hera turned her into a crane. It is also from this root word that we know geranium flowers, or cranesbills, which have a long, beak-like tip on their fruits.

The Gruidae are one of a dozen bird families comprising the avian order Gruiformes, meaning "crane-like." Gruiforms are an astonishingly diverse group with global distribution that includes approximately 190 species of cranes, rails, gallinules, coots, seriamas, trumpeters and bustards, and a handful of assorted sunbitterns, finfoots and mesites and the kagu, sungrebe and limpkin. Defining this group has been difficult and a multitude of classifications based on plumage, osteology (bones), DNA and behavior have not entirely resolved relationships among them. Most frequently, more aberrant group members, such as button quails and the plains-wanderer, have been classified among shorebirds in the order Charadriiformes. Nonetheless, many researchers consider the remaining gruiforms to be monophyletic; in other words, they all descended from a common ancestor that lived many millions of years ago.

Cranes themselves are probably most closely related to the limpkin (family Aramidae), a relatively large North and Central American marsh bird that feeds almost exclusively on apple snails (various *Pomacea* species), which it opens skillfully with the twisted tip of its bill, leaving behind telltale piles of unbroken shells. Although limpkins outwardly recall oversized, long-necked rails with their olive-brown mottled plumage, they have a crane's skeleton. The resemblance is so uncanny that many ornithologists consider limpkins to be primitive cranes, not unlike the crowned cranes of Africa.

This next-of-kin connection, also known as a sister relationship, between cranes and the limpkin is further substantiated by a fossil mosaic — one that exhibits characteristics of both bird groups. *Parvigrus pohli* lived in what

Many ornithologists believe that the limpkin — a somber-plumaged marsh bird with a penchant for eating apple snails — is the closest living relative of cranes.

is now southern France about 30 million years ago. This species was, indeed, a crane-limpkin cross although its smaller size and shorter bill and legs were more rail-like in proportions. Paleontologists believe that *Parvigrus pohli* was an offshoot of the gruid lineage that occurred before cranes evolved their characteristic long-legged, long-billed appearance. They suggest that this significant size increase in cranes occurred later as they adapted to foraging in more abundant grasslands and marshlands; forested environments were replaced by more open habitat types as the Northern Hemisphere became cooler and drier during the Oligocene and Miocene periods (10 to 30 million years ago).

The Gruidae are typically split into two subfamilies according to fundamental differences between some species. The most primitive cranes — two species of crowned cranes (genus *Balearica*) — are classified in the subfamily Balearicinae. Crowned cranes have distinctively fluffy body plumage and luxuriant golden feather crowns on their heads. They also have perching feet with long toes, and are the only cranes that can roost in trees. The remaining 13 species are attributed to the subfamily Gruinae. Unlike crowned cranes, the longer-billed gruine cranes have loud, resonant calls that result from an unusual confirmation of their trachea, or windpipe.

Gruine cranes (subfamily Gruinae) are further subdivided into three genera. The demoiselle crane and blue crane (genus *Anthropoides*) live primarily in upland habitats. They are smaller than other cranes, with shortened bills, which are well adapted for foraging on seeds and insects, and shorter toes that expedite running in open grassland habitats. Wattled cranes (genus *Bugeranus*) are much larger than demoiselle and blue cranes, and are somewhat more dependent on wetlands. Wattled cranes sport a pair of white wattles at their throat that increase in size with aggression. The remaining 10 species comprise the nominate genus *Grus*. They all exhibit adaptations to a fundamentally aquatic lifestyle, and many species are highly dependent on wetland habitats for nesting. Furthermore, their elongated necks and bills, long bare legs and large feet are useful for wading in shallow water or walking on soft sand or mud as they probe or excavate for food. Phylogenies, or evolutionary trees, of *Grus* cranes based on osteology, DNA and behavior recognize two closely related species groups within the genus: the Antigone group, comprising sarus, brolga and white-naped cranes, and the Americana group, which includes hooded, Eurasian, black-necked, whooping and red-crowned cranes. The relationship among the Siberian crane, sandhill crane and these eight species is not well understood.

The origin and early evolution of modern birds are shrouded in mystery. The traditional view, based primarily on fossils, suggests that modern birds arose from relatively few ancestral lineages that lived during the Mesozoic era — the time of the dinosaurs. This hypothesis also insists that most ancient bird species were devastated 65 million years ago in the mass extinction event that erased dinosaurs, pterosaurs, plesiosaurs and other wondrous creatures from the surface of our planet. Those few avian species that survived subsequently underwent rapid radiation to produce all modern bird orders within a 5- to 10-million-year period. Other paleornithologists believe many modern bird orders had their origins deep within the Cretaceous period of the Mesozoic era, and that they somehow snuck through the desolation that ended the reign of the dinosaurs. These studies use DNA sequences to predict the age of specific lineages but, despite compelling results, there are few Mesozoic bird fossils to support their conclusions.

Cranes are no exception: Mesozoic gruiform fossils are scarce. There exists an undetermined fossil species with crane-like characteristics, known only from the end of one hand bone (or carpometacarpus), which dates from about 70 million years ago. An earlier fossil species — the 100-million-year-old *Horezmavis eocretacea* — was once attributed to gruiforms but is now considered too generalized to classify within any modern bird order. Such is also the case with a modicum of other ancient fossils that danced briefly among the Gruiformes only to be translocated to other branches of the avian evolutionary tree. Did ancestral cranes walk among the dinosaurs? Unfortunately, the truth remains just beyond our grasp.

The extinction event that ended the Mesozoic era left a world impoverished of land animals and, whether or not the avian evolutionary bottleneck that occurred at this boundary was large or small, the survivors of the catastrophe became the founders of modern bird lineages. Pterosaurs, flying reptiles of the Mesozoic, no longer ruled the skies, and mammals had not yet left their mark on the planet. Birds, on the other hand, thrived in the equitable climates that stretched from equator to poles, and they diversified explosively to fill niches vacated by the dinosaurian extinctions. It was during these early periods of the new Cenozoic era that the ancestors of cranes made a brief play for supremacy.

The gruiforms that arose from the ashes of the Mesozoic stood 10 feet (3 m) tall with skulls the size of horses' heads. These swift, agile predators had powerful hooked beaks adapted to tear flesh and crush bones of the mammalian diet that they favored. Although their wings were not large enough to lift the monsters aloft, they were well clawed and sufficiently robust to subdue their struggling prey. They were the phorusrhacids, the "terror birds," that held the honorable position of top predator for many millions of years.

South America had been isolated from other continental landmasses for 100 million years, providing an optimal, competitor-free environment where the terror birds could become truly gigantic. Although their smaller relatives lived elsewhere in the world, including Europe and Antarctica, the largest phorusrhacids, such as *Phorusrhacos longissimus* and *Titanis walleri*, thrived in open grassland habitats from Brazil to Patagonia. Unfortunately, geographic isolation is forever fleeting as evolution and ecology bend with the influences of plate tectonics and global climate change. Thus, the reign of the terror birds ended only 2.5 million years ago — almost yesterday in geological time — with a massive dispersal event known as the Great American Interchange. The oncoming Pleistocene ice age caused huge quantities of seawater to be bound in polar ice caps that, in conjunction with uplift of the northern Andes Mountains, caused a 160-foot (50 m) drop in global sea level that produced the Panamanian land bridge. Across this narrow strip of land joining North and South America flooded throngs of animal species in search of underexploited habitats. South America's giant predatory birds were driven to extinction when large mammalian carnivores, such as saber-tooth cats, invaded their range from North America and outcompeted them for food. The terror birds may be gone but they leave behind living relatives, the seriamas — long-legged, predatory crane cousins — which still roam the pampas of South America in search of elusive prey.

The Eocene world — 33 to 55 million years ago — was peppered with other gruiform species. The geranoidids, known primarily from fossil leg bones,

were large, crane-like birds from North America. This group may have been the original stock for many early gruid lineages as they dispersed into the Old World across the Bering land bridge that connected North America with Asia. Some of their descendants became increasingly cursorial. Eogruids and ergilornithids were also large, crane-like birds; they thrived on the steppes and grassy plains of Mongolia and western Asia at least 20 to 30 million years ago. Many were flightless, or nearly flightless, and their powerful feet had flattened toes, which were well adapted to a fast-running life style. Similar to today's ostriches, some ergilornithids were didactylous, or two-toed. The reduction of toes is a

Titanis walleri, a giant flightless carnivorous crane cousin, terrorized the early horse *Hipparion* and other grassland mammals some 3 to 5 million years ago.

common convergent characteristic of cursorial animals — a horse's hoof is its last remaining toe, the others having been lost to evolution as the animal became optimized for locomotory efficiency.

True cranes may have descended from one of these deep Paleogene radiations; however, the fossil record is not about to reveal its secrets. Although there are many gruid species described, most are known to us only from a few bones, sometimes just a single bony element. We do know that crowned cranes, not unlike those that presently frequent Africa's marshlands and savannas, first appeared almost 50 million years ago. Since that time, at least 11 different species have graced the fossil record of Europe, Asia and North America. These birds are distinctive and primitive among living gruids, so it is not unimaginable that the earliest true cranes also wore golden crowns.

Ancient crowned cranes undoubtedly had a much larger geographic distribution than do today's species. *Geranopsis* occurred in England, *Eobalearica* in Europe and Asia. Perhaps the most intriguing crowned crane, however, was found in the Ashfall fossil beds in northeastern Nebraska. Twelve million years ago, a volcanic eruption in southwestern Idaho released a thick cloud of abrasive ash that choked the lungs of animals residing 1,000 miles (1,600 km) away; Ashfall fauna died within weeks of the eruption. Among these species was *Balearica exigua*, a small crowned crane that perished in association with a myriad of extraordinary late Miocene period animals, including the barrel-bodied rhinoceros (*Teleoceras major*), five species of horses, three species of camels, the saber-tooth deer (*Longirostromeryx wellsi*) and a small horned mammal of the genus *Mylagaulus* — a faunal assemblage very much like a present-day African savanna community, jackalope excluded. Hence, we can only conclude that crowned crane ecology has changed little in the past 10 million years. Extant crowned cranes are not tolerant of cold temperatures as are other crane species. So as the earth's climate cooled through the Cenozoic, crowned cranes living at more northerly latitudes became extinct, leaving just the two African species that we admire today.

Typical cranes appeared in the fossil record much more recently than their crowned kin; most fossil species date from the past few million years. At least seven different species of extinct cranes (mostly genera *Grus* and *Baeopteryx*) are known to have waded through wetlands and strolled across late Cenozoic grasslands. Fossil species demonstrate the wide geographic distribution that modern species do today, being found in most corners of the world except South America and Antarctica. Even Bermuda had an endemic (found only in) crane (*Grus latipes*) some 150,000 years ago. But not all crane species are relative newcomers to the earth. In 1928, ornithologist Alexander Wetmore discovered a fossil humerus (upper forelimb bone) from the Upper Snake Creek formation in Nebraska that is indistinguishable from modern sandhill cranes. This fossil stratum dates from the mid-Miocene (9 to 10 million years ago). Is this the oldest known fossil of a living bird species? The Upper Snake

Creek formation is not so distant from the Platte River, where some 400,000 sandhill cranes congregate annually during spring migration. Perchance sandhill cranes did dance among their compatriots some eight million years before humans walked this earth.

Anatomy and Physiology

Cranes are among the most statuesque of birds. The sarus crane stands almost 6 feet (180 cm) high; thus it is bestowed the honor of tallest flying bird. Some species tip the scales in excess of 25 pounds (12 kg), and are among the heaviest migratory birds. Large wings are requisite for carrying these considerable birds aloft; red-crowned crane wings measure over 8 feet (250 cm) between the tips of their white primary wing feathers. Even the smallest crane, the diminutive and pretty demoiselle crane, is taller than a Canada goose and heavier than a great blue heron.

Although all cranes are generally plumed primarily in shades of black, white and gray, their specific and characteristic colors reflect their adaptation to the breeding habitat that they use most frequently for rearing their young. For example, the larger species, such as Siberian, whooping and red-crowned cranes, which live in immense open wetland areas, are predominantly white or have bold white signal patches. Signal patches are bright or eye-catching plumage markings used for visual communication during displays. These large, imposing cranes wish to seen because their conspicuous appearance helps to advertise occupation of their nesting territory. There is little concern if their high visibility attracts predators; most predators are unlikely to challenge directly the aggressive side of a breeding crane. Also, many cranes that nest in open habitats have evolved distraction displays, including feigned broken wings, that are used to lure a potential predator away from the nest. Here again, being conspicuous is of considerable value.

On the other hand, the smaller crane species such as hooded cranes, which inhabit more forested areas or small isolated wetlands with limited visibility, benefit from being camouflaged against predators. These species are typically clothed in shades of gray. Nesting Eurasian cranes and sandhill cranes also use their bills to "paint" their feathers with rusty brown iron-rich mud to make themselves less conspicuous.

Most cranes have bare, bright-colored (usually red) skin on their heads or face. Skin patches are brightest during breeding season; they become even more intense in color when heightened excitation associated with a potential threat causes the thin skin to be engorged with blood. Cranes that sport these bright colors often flash their red patches by bowing their heads when defending their territories against rivals or intruders. In addition to red cheek patches, crowned cranes have elaborate golden feather crests on their heads. Mated pairs of crowned cranes regularly preen each other's crests, perhaps as a sign of devotion

Distinctive differences in plumage coloration make the adult whooping crane (foreground) purposely more conspicuous in appearance, and its offspring (background) much less so.

to their lifelong pair bond. Demoiselle and blue cranes lack red skin patches and have fully feathered heads; however, they can erect plumes on the sides of their head during display, giving them a somewhat cobra-like appearance.

The plumage of young cranes is unlike that of adults; in general, they are designed to be missed. Juveniles are typically reddish brown or gray to make them less visible to predators. Cryptic young are particularly important in large cranes, such as whooping cranes, that are predominately white as adults. These birds retain some of their rusty cinnamon plumage into their second year of life, giving them a mottled appearance that provides a degree of camouflage. However, by age two, when the first notions of romance enter their heads, they look entirely like adults.

Birds must keep their plumage in impeccable condition. Not only are serviceable feathers mandatory for efficient flight, they also provide insulation from heat or cold and are critical elements on exhibit during territorial and courtship displays. Feathers cannot be repaired once fully grown, so they must be replaced when worn. Feather replacement (molt) typically occurs in a predictable sequence; most cranes undergo their annual molt in summer after nesting is finished. This allows them to restock their energy reserves, which were depleted during egg laying and incubation, when food is plentiful. Wing flight feathers are lost almost simultaneously in some species, rendering them virtually flightless for a few weeks. Fortunately, new feathers quickly replace the gaps in their wings, putting them in good stead for their upcoming migration.

Cranes are expert fliers. They have large, broad wings, strongly curved in profile, that are called slotted high-lift wings. Lift is the force required to keep a bird aloft; the larger the bird, the more lift is needed to overcome the downward pull of gravity. Lift is generated as air passing over the wing's longer top surface (because of the curve) travels faster than air passing under the wing. This creates a low-pressure region above the wing and a high-pressure region below the wing, the net effect being that the wing rises. Cranes and other large birds, such as pelicans, herons and eagles, must generate substantial lift in order to fly. Evolution accomplished this by equipping them with large wings that have a high surface area (length × width) on which to provide adequate lift. In addition, they have wing slots — spaces between the finger-like primary flight feathers (or primaries) located at their wing tips. As the slots open, each primary acts like a miniature wing in generating a bit more lift. Aircraft also create wing slots to increase lift while flying at slow speeds by extending slats at the wings' leading edges. Slotted high-

lift wings are also useful to birds in this regard because the ability to fly at slow speeds without stalling allows a large bird to soar — to fly without flapping. This energy-efficient means of travel is particularly useful during long migrations.

Many soaring birds such as cranes have broad wings tipped with finger-like feathers that separate to form slots; slotting creates additional lift needed to keep these large birds aloft.

All cranes, except red-crowned cranes, have blackish primaries; dark feather tips are particularly common among migratory birds. The color is produced by melanin, a biochrome pigment that is responsible for black, brown and gray coloration in most animals. Melanin is very resistant to wear, so a feather tipped with black is much stronger. Primary feather tips suffer considerable wear because a vortex of turbulence is formed there when the bird flaps its wings to generate forward thrust. Dark wing tips also serve as signal patches, particularly when they contrast strongly with pale body plumage, and can add a dramatic element as wings are raised and lowered during display.

All volant birds use their secondary wing feathers — which originate on the ulna, the largest bone between wrist and elbow — to generate lift. Cranes, however, have exceptionally long and curved secondaries. In some species, such as Eurasian and black-necked cranes, the secondaries extend backward and cover the shorter tail to create a "bustle" not unlike the small pad worn by late-19th-century women to puff out their skirts. Elongated inner secondaries of blue, demoiselle and wattled cranes trail almost to the ground when the bird is standing at rest with folded wings, and are often mistaken for a long tail. Cranes often raise their secondaries with a flourish as they call in unison or leap high into the air as they dance. Furthermore, extended wings create an imposing sight when a territorial crane encounters a potential intruder.

The internal anatomy of cranes would be unremarkable if it were not for one notable exception: many species bear the equivalent of a bassoon or a double bass beneath their feathery breasts. Birds do not have vocal chords. Instead, they produce sounds using an organ called a syrinx that is located in their chest at the junction of the trachea and two primary bronchi, which lead from the lungs. The syrinx produces sound as vibrating air is forced from the lungs through narrow syringeal passageways; these passageways can be changed in shape and size by muscles to produce the variety of notes and tones that we associate with birdsong.

It has long been known that tracheal structure is a significant contributing factor in the production of the characteristic loud, trumpeting calls exhibited by many crane species; the complete anatomy of Eurasian crane vocal apparatus was described by 1575. We also know that air passing through a long column produces a lower tone. This phenomenon is used in the design of many musical instruments; incrementally closing the holes of a flute — effectively lengthening the column through which air passes — produces increasingly lower notes. The longer trachea also increases resonance, particularly when its coils are embedded in a large radiating structure such as the sternum (breastbone). The avian sternum is the attachment site for major flight muscles and, consequently, is much larger than mammalian sterna. Tracheal coils act like a violin's bridge to transmit a tiny sound source to the large resonating chamber, resulting in highly amplified calls that can be heard for 2 to 3 miles (3 to 5 km).

Elongation of the trachea is most pronounced in typical *Grus* cranes, which produce very loud, penetrating bugling calls. The extent of intrasternal coiling varies considerably, however, ranging from simple looping in sandhill cranes to immense double looping in whooping cranes. The total tracheal length ranges from about 24 inches (60 cm) in sandhill cranes to 60 to 65 inches (150 to 165 cm) in whooping cranes and red-crowned cranes, and as tracheal length increases, the calls of adult birds become increasingly more "whooping." Young birds do not whoop until they are several months of age. Prior to this, the forward end of the sternum is still cartilaginous, which allows the tracheal loops to gradually penetrate this soft region of the breastbone. As the trachea elongates through development, the space inside the sternum grows to accommodate it and, finally, the sternum becomes ossified (bony) to produce the resonating chamber.

Other non-*Grus* cranes, such as demoiselle, blue and wattled cranes, have less extensive intrasternal coiling; the tracheal loop is a simple S-shape. Although these species can produce relatively loud calls, they are not characteristically "whooping," but have been described as guttural, shrill or screaming in nature.

The trachea of crowned cranes is not coiled at all — resembling that of limpkins — and passes from the lungs directly between the branches of the furcula (or wishbone) to the neck, with a total length of about 20 inches (50 cm). A similar tracheal condition was found in the fossil crowned crane *Balearica exigua*, from the late Miocene. Crowned cranes produce goose-like honking notes. They are, however, capable of uttering low-pitched booming calls when their throat sacs, which serve as resonating chambers, are inflated.

The white-naped crane's long neck encloses an equally long trachea (windpipe) that functions like the coiled tubes of a bassoon to produce loud, trumpeting calls characteristic of some species.

Feeding Ecology

All cranes are omnivorous — they eat a variety of plant and animal foods — which they seek in wetlands, croplands, native grasslands and fallow fields. They typically choose their food opportunistically, readily devouring items that are abundant and easy to procure. All cranes are somewhat generalistic in their feeding habits although a few species are pickier than others. Also, chicks typically consume proportionately more animal material than plant material because they need added protein for growth, which can be astonishingly rapid; sarus crane chicks are 4 feet (120 cm) tall at about three months of age.

Favored plant foods include the tubers, roots, stems, shoots and seeds of submergent and emergent aquatic plants in wetland habitats; seeds, leaves, acorns, nuts, berries and other fruits found in upland areas; and crop produce, including waste grain (barley, oats, sorghum and wheat) and corn, sweet potatoes, melons and alfalfa. Cranes consume a diversity of animal foods, such as fishes, frogs, beetles, insects, worms, snails, crustaceans, small rodents, reptiles (snakes and lizards) and occasionally small birds or their eggs.

Long legs enable cranes to wade through water and tall grass or reeds as they forage, and they use their long bills for picking, probing and digging for food items. Grains and corn are plucked from the ground or from stalks or ears lying in fields. Cranes search for worms and other invertebrates with their head held down, enthusiastically probing in mud, sand, vegetation and even dried cow droppings with their bills. Larger prey items, such as frogs, fish and crabs, are actively stalked. Unlike herons, which stand like statues waiting for suitable prey to pass by, cranes move in circles or walk in single file along wetland margins as they seek their food. Digging cranes are somewhat more stationary, often excavating the same hole relentlessly until all the interconnected buried tubers or roots are discovered.

Cranes usually feed for extended periods in the morning, rest during the heat of the day, and then return to feeding grounds in late afternoon; typically one-half to three-quarters of daylight hours are spent in search of food. They often forage where vegetation is fairly low and visibility is unrestricted, which allows them to keep watch for predators and competitors alike. During breeding season, most species feed alone or in pairs and family groups; territorial adults vigorously drive off other cranes that attempt to feed in their vicinity. However, they are generally more accommodating during migration and in winter. In particular, species that feed primarily on vegetable matter when not breeding, such as sandhill, white-naped and Siberian cranes, may forage quite peacefully in flocks. Animal-eaters, including whooping and red-crowned cranes, tend to exhibit some territorial behavior even in winter.

Despite these generalizations, foraging behavior varies somewhat among cranes, which is, of course, reflected in anatomical differences. Shorter-billed species generally feed in drier uplands habitats, such as grasslands and savanna.

Though blue and demoiselle cranes occasionally dig, they typically pluck their food from the ground. Crowned cranes, which have the shortest bills of all, enhance this process by stamping the ground with their large feet to scare up insects that may be lurking there. They also graze on vegetation in the manner of geese.

Longer-billed cranes typically feed in wetlands. The longest-billed species — sarus, wattled and brolga cranes — are diggers that use their powerful bills to excavate buried tubers and roots from heavy, waterlogged soils. Whooping cranes and Siberian cranes use their long bills to probe muddy bottoms of shallow wetlands, usually in water 10 to 25 inches (25 to 70 cm) deep, for small fish and crustaceans. Eurasian cranes, however, are more generalistic, and find food in shallow water, low vegetation or beneath the ground.

Nevertheless, studies of crane feeding ecology indicate that there is flexibility in foraging strategies that, in part, echoes their opportunistic approach to survival. This is, perhaps, easiest to see where more than one species occur at once. For example, four crane species coexist in winter at Poyang Lake in China by partitioning available foraging habitats. Siberian cranes feed on mudflats and in shallow water; white-naped cranes use wetland margins;

Grey crowned cranes forage among the dry savanna grasses by plucking seeds with their short bills and stamping their feet to dislodge hidden insects.

hooded cranes feed in croplands and meadows adjacent to the wetland areas; and Eurasian cranes, the generalists, occupy the spaces in between the others. Siberian cranes wintering with sarus cranes in Keoladeo National Park in India, however, feed on sedge tubers in deep water while the sarus cranes use shallow water areas. In Australia, where sarus cranes coexist with brolgas, they are relegated to feed in drier, less productive habitats, while the brolgas forage in lowland sedge marshes. Perhaps the redistribution of some species into less equitable feeding habitats is spawned by competition rather than cooperation. Cranes are differentially aggressive, and when two species encounter each other, one species typically wins most of the battles. When push comes to shove, sarus cranes tend to dominate Siberian cranes, just as white-naped cranes will prevail over hooded cranes.

Many cranes rely on agricultural fields during their annual cycle, often nesting or roosting in wetlands adjacent to croplands in which they forage. They have little impact on farm production if they feed primarily on gleanings, or leftover grain after harvest, during winter and at early migratory stopovers, both of which usually occur outside of the growing season. Gorging on waste grains is an energy-efficient way of obtaining food, so they are eaten readily during migration. Some species, including sandhill cranes staging on the Platte River in March, survive almost entirely on waste corn in nearby fields. Elsewhere in the world, however, cranes can cause significant damage as they probe for newly planted seeds or tender seedlings. Migratory flocks are particularly devastating during fall harvest or in spring when crops are germinating. Chasing ravenous cranes away from fields has no more than a temporary effect; they just move to another field and then return when the disturbance subsides. These unfortunate events have contributed to widespread persecution in some African and Eurasian countries, and demoiselle, blue, hooded and black-necked cranes are among the species that are succumbing to intentional poisoning and shooting. Although it is rarely implemented, farmers have successful lured hungry birds away from commercial crops by enticing them to feed on inexpensive foods planted near roosting areas. Perhaps additional incentive and compensation programs would encourage farmers to live peaceably with cranes.

On the other hand, provision of grain has played a critical role in some species' survival. Artificial feeding began in 1952 in southeastern Hokkaido in Japan, where about 30 red-crowned cranes wintered along streams and hot springs. A severely cold winter froze the open water where they fed. Fortunately, the hungry cranes readily accepted corn scattered on the ice by school-children. Japan is now home to three major feeding stations, and several smaller ones, that feed more than 600 red-crowned cranes — second only to whooping cranes in their rarity. The Izumi feeding station on the island of Kyushu supports 80 percent of the global population of hooded cranes, 40 percent of white-naped cranes and 35 percent of red-crowned cranes. Artificial feeding has fostered rapid population growth and, hence, initiated partial recovery of these endangered and vulnerable species. Regrettably, artificial feeding is not without risk, because it causes birds to concentrate in high numbers, where they are prone to a single catastrophic incident, such as disease outbreak, unintentional poisoning through food contamination, or an inclement weather event.

Behavior and Communication

Cranes perform a multitude of behaviors that include both individualistic activities to enhance survival and visual and vocal communication signals directed intraspecifically, which serve to perpetuate the species. Cranes are basically diurnal; they forage, tend to their young and socialize with one another during the day. They also spend a substantial part of their day preening, particularly when molting. Feathers must be straightened and cleaned so that they remain in good working order. Birds lavish an inordinate amount of time on their flight feathers (wing and tail feathers), first nibbling at the base of each feather and then drawing it through their bill from base to tip. They are, in effect, "zipping" the feather up. Flight feathers, by design, are both flexible and strong. They have a central shaft, called the rachis, that supports a broad, flat vane on each side. The vane on the feather's leading edge is narrow, the trailing edge wide, so that each feather acts as a miniature airfoil. The vanes themselves are not solid but are composed of innumerable lateral branches called barbs. Each barb also has a central axis that bears even smaller barbules, which are tipped with tiny hooks. These hooks extend to a neighboring barb, where they grasp and slide along other barbules thus providing flexibility while maintaining strength. As a bird runs its bill along the feather from its base, it is re-engaging any hooks that may have become detached, thus restoring optimal aerodynamic function.

Cranes typically rest at night. During the breeding season, they stay near their nest, where they are ever watchful, either incubating or brooding their young or standing guard against predators. When not breeding, most will roost in flocks at traditional sites. Crowned cranes frequently roost in trees; however, other species prefer to roost in shallow water or on sandbars or mudflats that are surrounded by water. Cranes usually rest and sleep while standing on one leg; they often switch legs throughout the night at intervals. Occasionally, they will sleep in a sitting position with folded legs beneath them, with their head placed on their breast or rotated and tucked into the shoulder feathers. When roosting, each crane usually keeps one "peck distance" from its nearest neighbor to maintain peace in the ranks. Flock behavior confers the benefit of vigilance — one crane may nap while another oversees the area. Should a flock member hear an unfamiliar sound or see potential danger, it will utter an alarm call to rouse other roosting birds. If the threat is genuine, they will flee on the wing.

Crane flight is characteristic. Typically, after a short running start, they leap from the ground and become airborne with a few powerful downstrokes of their large wings. Unlike herons and egrets, which retract their necks into

S-shaped loops when they fly, cranes hold their necks outstretched. Their long legs and feet trail gracefully behind them, acting as a counterbalance to their extended heads and bills. Cranes also beat their wings with unique style. They have a slow, deliberate, strong downstroke that provides sufficient forward thrust to overcome drag generated by their large bodies, and a quick upstroke (recovery stroke) that efficiently returns the wing to its former position. This snapping rhythm of crane wings in flight is best observed as they gain altitude quickly.

Cranes conserve energy while flying by soaring almost effortlessly on their broad, outstretched wings.

Powered, flapping flight consumes considerable energy, however, particularly for such a large bird. Consequently, most cranes soar on their broad, slotted high-lift wings when conditions are favorable. Their wings are particularly well designed for soaring on thermals, rising columns of warm air created as the sun heats the earth's surface. Soaring cranes can remain aloft for hours, spiraling slowly upward on set wings in these rotating currents. Some species, such as the Eurasian crane, routinely climb to altitudes in excess of 6,500 feet (2,000 m) in thermals. Whooping cranes often soar a mile (1.6 km) high above the plains of the southern and central United States. Soaring is also an effective way of escaping the day's heat, because air temperatures become increasingly cooler at higher altitudes; cranes often soar for no other reason than this.

And as we see strainge Craine
are wont to doe
First stalke a while ere they their
wings can finde,
Then soare from ground not past
a yard or two,
Till in their wings they gathered
have the winde;
At last they mount the very
cloudes unto,
Trianglewise according to their
kind.
— Ludovico Ariosto —

The diverse array of social behaviors exhibited by cranes varies considerably throughout their annual cycle. During breeding season, which lasts three to five months, mated pairs and their offspring occupy a territory, a defended space that they use for their daily activities: nesting, feeding and roosting. They are typically more gregarious during the nonbreeding season, and most migratory species travel and winter in loose flocks. Many nonmigratory cranes also occur in nomadic flocks that make local flights during the nonbreeding season in search of secure roosts and plentiful food. Depending on the species, flocks of cranes can be large or small, and may comprise family groups or unrelated members. Social interactions in the nonbreeding season can be violent, but they are not intended for defending space. Instead, they serve to establish and maintain dominance hierarchy in the group, to introduce young birds into the flock and to allow individuals to assess the potential for future mates. In all of their social and communicative activities, one bird is the sender of the message and the other the receiver. Regardless of whether or not the intent is aggression, appeasement, altruism or sexual or familial interaction, the signals used in communication can be vocal, visual or a combination of both.

Notable environmentalist and philosopher Aldo Leopold once described an approaching flock of sandhill cranes as first a "tinkling of little bells," then a "baying of some sweet-throated hounds" and finally a "pandemonium of trumpets, rattles, croaks and cries." Indeed, cranes produce everything from quiet contact quacks to loud, bugling unison calls although the volume and tone vary substantially between species. Like other birds, cranes have a vocal repertoire that contains a set of characteristic elements that, when used alone, or in combination, convey specific information according to situation. The information content and nature of the communication adjust as birds age because of differential development of their vocal apparatus and their changing needs, those perceived as critically important to that age cohort.

Young cranes begin vocalizing before they hatch. Sandhill cranes utter a constant, trilling call for 10 to 39 hours before they first pip (crack) the eggshell. Their parents respond to this call by purring to the eggs to initiate contact with their chicks inside. This parental response is the first step in the imprinting process, whereby chicks learn to recognize the individuals that are their parents. Imprinting is an important survival strategy because it establishes the parent as an object of "habitual trust" on which the chick may rely for food, protection, guidance and education.

Crane chicks produce five characteristic calls from hatching until almost one year of age. They utter high-pitched peeping notes to beg food from their parents. They also give a low, purring contact call when they are content and in their parents' company. This call is uttered almost continuously as the family forages together. However, when the chick is cold, hungry or temporarily separated from its parents, it produces a loud, insistent stress call that immediately draws any parent within earshot to its side. When the young crane fledges, usually at age three to four months, it begins to give a flight-intention call, which is a brief high-frequency, unbroken call uttered as the bird stands facing into the wind; birds typically take off into the wind because their wings produce more lift when the speed of air passing over them is increased. Crane fledglings also give an alarm call — low-pitched, rapid, broken notes — in response to a perceived threat or stressful situation.

Peeping sounds of young cranes give way to the deeper-toned, louder adult voice by nine to eleven months of age. Begging and stress calls disappear; contact, flight-intention and alarm calls are retained but they become lower-pitched. These immature birds produce two new calls: the plaintive location call, a loud, screaming syllable used to locate other cranes that are no longer visible, and the guard call, which is given as a threat to other members of the species.

At about two to three years of age, cranes begin to seek out potential mates. They learn a precopulatory call, a series of purring notes, which is given by paired birds just prior to mating. The most significant call to emerge at this time, however, is the unison call. Characteristic of species, the unison call is a long, complex, coordinated duet between male and female cranes that is combined with specific postures and placement. The call varies from several seconds in duration to over one minute, and is repeated regularly throughout the day. Both males and females participate in the duet with their own unique contribution of voice and behavior.

Unison calls are most commonly heard just before breeding season, and may be used to announce occupation of territory, or in courtship rituals to establish and maintain the pair bond. Cranes typically mate for life. The unison call helps partners come into breeding condition simultaneously, and it may be particularly important in the timing of ovarian function in females. New couples that are less familiar with their partners' idiosyncrasies may call frequently, but their calls will be less well coordinated than those of experienced pairs. The longer the pair is together, the more elaborate and precise their duet becomes.

In all cranes, the unison call is uttered as the bird holds an erect, alert posture; however, the position of the calling bird's head and wings varies with species and gender. The head may be held forward, vertically or thrown back; the wings may be at rest, lifted slightly or raised in a flourish above the bird's back. Crowned cranes begin their minute-long unison call by lowering their head to shoulder level, inflating gular sacs at their chin, and uttering a series of honks followed by a long array of booming calls. Unison calls are generally shorter in duration in other cranes; also, they lack booming notes but are louder and more penetrating in nature. These birds usually stand side by side while they call and, although the participants call more or less in synchrony, their gender can be determined by their voice tone and the posture that they assume while vocalizing. Females initiate the call in many species, and their voices are somewhat higher-pitched than the males' contribution. Also, females often give two or three shorter notes for each single response note of the male. Female calling posture is often less flamboyant than the male's, and is usually restricted to a raised or reclined head.

All cranes dance, and although this behavior is, indeed, ancient in origin, its evolution remains cryptic. A few other crane-like bird species, including trumpeters (family Psophidae) and egrets (family Ardeidae), do perform some sort of courtship dance but their behaviors are not so elaborate, habitual or frequent, nor are they so characteristic of species. Limpkins — cranes' closest

Opposite: Nesting cranes are devoted parents that ensure the well-being of their unhatched chicks by rotating the eggs to maintain an appropriate incubation temperature for proper development.

living relatives — do not dance at all, and their behavioral similarities are restricted primarily to their threat and appeasement displays.

Dancing in cranes, however, is systemic, addictive and all-encompassing. The long, intricate, coordinated behaviors include at least seven distinct steps: (1) jumping and leaping; (2) bowing and head bobbing; (3) wing spreading and flapping; (4) whirling; (5) leg dangling; (6) neck weaving; and (7) billing. Many species dash about, sometimes in circles, between other dance moves. Cranes often fling objects, such as sticks, moss grass, and feathers, into the air with their bill during the dance, and kick or peck at the debris as it falls. Dancing may also be accompanied with vocal display.

Dance form varies somewhat between species groups. For example, crowned cranes bow less than other species, but head bob more. Smaller cranes, such as demoiselle and blue cranes, are often more energetic than larger species; adult blue cranes can dance incessantly for several hours. Demoiselle crane performances have been described as "ballet-like" because they contain much less exaggerated jumping than the more deliberate, high, wing-flapping leaps that characterize dances of the larger *Grus* cranes, such as whooping and red-crowned cranes.

Dance components often become more intricate as cranes age. Flightless chicks typically dance by flapping their short wings rapidly while bouncing up and down and running about. As they get older, they incorporate more elaborate steps, including bowing at 14 weeks of age, and stick- or grass-tossing at age 17 weeks. Ten-month-old red-crowned cranes have been known to dance all day around their parents and other birds in their flock. Thus, dancing may be an individual or a group activity. Pairs are most frequently observed dancing, but not always together; occasionally one pair member dances around its mate for several minutes to entice it to join the display. Its mate may do so; however, sometimes it may be entirely indifferent, or even annoyed at the invitation, which is indicated by a few infuriated pecks at the dancing bird as its passes near. Nevertheless, dancing can be contagious. On occasion, a crane may spontaneously break into dance, which then spreads like wildfire through the flock. In the resulting mêlée, some dancers appear more synchronized in their contributions, while others merely indulge themselves in crane frenzy.

Dancing was once considered to be merely courtship display because paired cranes participate in variably synchronized dances to establish or renew their pair bond. Studies have demonstrated that newly formed pairs dance more than experienced couples, likely because the need to thwart undue aggression between individuals and synchronize readiness to mate will decrease the longer the birds know each other. However, the suggestion that dancing functions solely for courtship diminishes the behavior's importance in crane society. Dancing occurs virtually throughout the year in many species, and is most commonly observed among unpaired cranes, two to three years in age, that have not yet selected a life mate. Thus, dancing may also provide some general measure against which social status, or willingness to participate in social activities, is assessed. Flock dancing — frequently observed in sandhill, Eurasian, sarus, brolga and red-crowned cranes in fall and winter — offers a venue in which young birds may choose a mate. Furthermore, older birds may use this opportunity to appraise their position in the flock's dominance hierarchy or to work out disputes among their colleagues. Cooperation among flock members is of particular importance in species that migrate, and a well-established and tested rank structure is critical for success. Cranes also dance to release tension when disturbed or angry, to thwart aggression, to distract predators and to advertise occupation of territory. However, one cannot dispute the fact that sometimes they dance for what appears to be no reason at all; captive cranes dance in fine weather, and penned birds dance when they are released from their enclosures. Even two-day-old cranes dance in play on their gawky long legs. Perhaps this is merely the reflection of a state of mind, or heart, that we as humans attribute all too frequently only to our own species: *la joie de vivre*, the joy of life.

Crane displays, however, are not all fun and games. Cranes are highly territorial, and will defend vigorously both nesting and feeding territories, as well as their mates and their own individual space. Agonistic, or combative, behavior is most frequently observed among males although it is not exclusive to them in some territorial encounters. Ethologists have recognized several territory types that are defended by cranes. The size and location of the protected space, and the degree of aggression demonstrated in its defense, vary considerably within and between species. Factors such as time of year, food availability, population density and the individual nature of the bird (or its humor at the time) can also influence agonistic reactions significantly.

Cranes typically uphold the five following territory types:

(1) Males will defend their individual space against all other males of their species. Rarely, a male will peck at his female if she approaches too closely as he feeds. This territory boundary will be small and move with the birds as they undertake their daily activities.

(2) A male crane also defends space around himself and his mate, vigorously protecting her from danger and jealously guarding her against the advances of other males.

(3) Males defend an optimal feeding space where preferred food is plentiful and relatively easy to procure. This is a territory with distinct, immovable boundaries because it provides specific habitat elements required for survival, such as wetlands, crop fields or mud flats.

(4) Both males and females will defend the patch of ground that they occupy while roosting in a flock. Although small in size, its specific location may be very significant. Roosts that are surrounded by water or positioned at the heart of the flock are less prone to predation and are, therefore, more highly valued.

When defending their territory or deterring predators, most species, including the sarus crane, adopt imposing postures in order to look larger and more menacing.

(5) Finally, male cranes lead, often assisted by females, in the defense of a very large territory that is used for nesting.

Crane breeding territories are frequently 1,000 acres (400 hectares) or more in size, and may be used year after year by the same mated pair. Once the space is initially acquired, occupying birds are given the "home field advantage" by other cranes; consequently, it takes considerably less effort to maintain territory boundaries than it did to first establish them. Typically, the nesting territory contains some foraging habitat, which is particularly important when chicks are newly hatched and unable to travel too far from the nest.

Many agonistic, or conflictive, encounters begin with ritualized behaviors that inform the participants of each other's intentions. Sometimes, these "pre-fight" postures are effective in diffusing the situation, allowing one bird to withdraw before a potentially injurious attack occurs. For example, 13 of

15 crane species have bare red skin on the crown of their head that is used ceremonially in agonistic displays. When the bird is stressed or agitated, this skin becomes engorged with blood, which enlarges the patch and makes it more brightly colored. A defending or attacking crane frequently orients its head so as to clearly show its opponent the state of its crown and, consequently, its mood. Threatening birds also raise and fan their large wings; sometimes, they stand facing (or walk toward) their opponent while quickly flapping their wings alternately. This violent side-to-side shaking makes the bird look bigger and more menacing. The male may also touch his lowered bill to his leg or flanks in exaggerated preening movements. The precise meaning of such ritualized behaviors is often difficult to interpret, and they would not be readily apparent if they were not incorporated into other, obviously threatening activities; sometimes messages embedded in animal behavior can only be inferred by looking at what came before and what response is elicited. An aggressive crane often performs a stiff "parade walk" — moving slowly around the object of its concern in a circular path, while tilting its bright red head toward center to make clear its intentions. Accompanying low-pitched snorts and growls add impact.

If these threatening displays are directed toward another crane, then the opponent may demonstrate an appropriate appeasement display to avoid a violent encounter. In effect, appeasement requires the opposite of threat. The bird maintains a horizontal body posture with its neck and bill retracted and wings folded at its sides; the red crown is diminished both in size and effect. Also, the bird moves away from its potential opponent in a rather loose-jointed manner that is distinctly different from the rigid, strutting parade walk. If appeasement is not effective, or if one party will not give way, an attack ensues. An attacking crane is a fearsome opponent as he runs rapidly toward the intruder with his wings flapping wildly. As he draws near, he thrusts himself into the air with his powerful legs and jumps forward, kicking violently with the sharp claws of his feet. If his challenger chooses not to flee, he will jump and kick his opponent until one bird finally concedes and withdraws beyond the area of dispute.

Reproduction

Cranes commonly live 20 to 30 years in the wild, and have exceeded 80 years in captivity. They produce few young every year — often only one. In order to perpetuate the species they must successfully raise young every year of their long lives. Hence, it is somehow fitting that all crane species are strictly monogamous, maintain lifelong pair bonds and produce young with a prolonged period of dependency on their parents. Most species begin to establish pair bonds when they are two or three years old, but usually do not breed successfully until age five or more. Pairs stay together throughout the years unless early attempts at breeding are unsuccessful, which occurs quite often, or there are repeated failed breeding seasons later in life. Under these circumstances, birds will choose another mate.

As nesting season ensues, there is increased activity among adults. Dancing and unison calling occur more frequently as experienced birds renew their pair bonds and new couples become better acquainted. Older birds seem more comfortable with each other — having passed this milepost many times already — and are more coordinated and relaxed in their activities. Cranes are territorial during nesting and will not breed unless they can secure, and successfully defend, an acceptable breeding territory. Pairs define their territory with unison calling, threat behavior and attack, if necessary. Both males and females participate in maintaining territory boundaries; however, most defense is done by the male, particularly after nesting has begun, when the female is more focused on nest construction, incubating eggs and tending young.

Many species use the same nesting territory from year to year. Consequently, if all potential nesting territories in an area are occupied, then younger individuals must wait until one is vacated, typically through the demise of its owner. Migratory cranes that breed in the northern temperate or arctic regions begin to establish nesting territories soon after they arrive from April to June. Onset of breeding in nonmigratory cranes, particularly those that live at subtropical and tropical latitudes, is more variable; however, it often occurs during the local rainy season.

Cranes' territories are usually quite large, dispersed and arranged along patches of preferred habitat; they rarely nest on territories of less than 60 acres (25 hectares). Some species, such as red-crowned cranes, have enormous territories that may measure some 1,000 to 3,000 acres (400 to 1,230 hectares) in area. Territory size also varies with the overall quality of available resources and proximity to potential disturbance. Indian sarus cranes have successfully raised young on abundantly resourced, secluded territories as small as 2.5 acres (1 hectare); nonetheless, this occurrence is quite unusual.

Most cranes establish their territories to include shallow wetland habitat that provides parents and newly hatched young with all resources needed for nesting and feeding. The degree of dependency on wetlands throughout the breeding cycle, however, does vary. Large white cranes, such as whooping, red-crowned and Siberian cranes, typically stay in wetlands from the onset of nesting until they depart for their wintering grounds, and feed their young primarily aquatic foods. Others, including brolgas and Eurasian cranes, nest in wetlands but lead their young to drier, upland areas to feed soon after the chicks hatch. A few species prefer to nest on dry ground or, sometimes, above it. Cuban sandhill cranes (but not other subspecies) and demoiselle cranes rear their young on dry ground, and grey crowned cranes occasionally nest in trees, often in the abandoned nests of other large birds.

Disturbance is of grave concern for many breeding cranes, particularly among young pairs. Early nesting attempts are often fruitless, so space and solitude are absolute requirements for new parents so that they can learn how to successfully hatch and rear their young. Even among experienced birds, disturbance can hamper nesting.

As a result, most cranes prefer to nest in open spaces where they have good visibility over a great distance so that they can see approaching predators or other potential intruders. In addition, cranes rarely nest within a few miles (several kilometers) of human activity. There are a few exceptions and, perhaps, these species may fare better in our already crowded world. Some sarus cranes can nest in human-modified environments, such as small village ponds, so long as they are not harmed. Also, demoiselle cranes, which traditionally nest in steppe habitat, will maintain their territories as the land is converted to agriculture, provided that farming activities are not overly disturbing at critical times in their breeding cycle.

Once ownership of territory has been well established, the pair chooses a specific site on which the nest will be built. They often cement their agreement with a little unison calling before construction begins. Species that nest in wetlands shape their nest from aquatic vegetation torn up from the surrounding area and mounded haphazardly in shallow water. They begin by shredding stems and leaves of sedges, cattails, bulrushes and other aquatic plants located near their proposed nest site and heaving them backward over their shoulders. As the birds gradually move toward the site, they continue to gather and throw debris within their reach. Eventually they have accumulated a sizable pile of nesting material forming a low, flat platform surrounded by a water moat. Sometimes prospective parents trample the center of the pile with their feet to make a hollow spot for the eggs. This nature of nest can take one day to one week to construct, depending on how much vegetation is amassed. Demoiselle and blue cranes, which often nest on dry ground, prepare little for the coming blessed event. They merely gather together a few pebbles or bits of vegetation to provide some camouflage for the eggs. Sometimes they just create a bare scrape in the ground to prevent the eggs from rolling away.

Egg laying is timed so hatching occurs when food is plentiful. In migratory species that breed at northern latitudes, this means spring. Spring hatching also allows time for maximum growing so that young will be large enough to leave for the wintering grounds when the weather deteriorates in autumn. Nonmigratory species that nest at lower latitudinal temperate regions, such as the Florida and Mississippi sandhill cranes, also nest in spring like their migratory relatives. However, they have a longer breeding season without the threat of inclement weather, and will frequently lay a second clutch of eggs if the first attempt is unsuccessful. Most tropical cranes lay their eggs during or just after the peak of rainy season, when food is plentiful, regardless of the month.

Female cranes lay one to four eggs, although, they most typically lay two eggs that are white, buff or blue in color and often spotted with brown or purple. Interestingly, the intensity of egg coloration varies with latitude. Species that nest in tropical or subtropical regions produce paler eggs than species that breed in colder climates, which lay more darkly pigmented eggs. This evolutionary adaptation lessens the likelihood that uncovered eggs will over-heated or overcooled by absorbing or reflecting the sun's heat. The size of crane eggs is more or less relative to the female's body size, ranging from about 2 × 3 inches (50 × 80 mm) in size and weighing 5 ounces (135 g) in demoiselle cranes to approximately 2.5 × 4 inches (65 × 100 mm) and 8.5 ounces (240 g) in sarus cranes.

Most cranes lay only one or two clutches of eggs each season, although tropical species occasionally raise three or four broods. Species that nest where the summer is short, such as whooping and Siberian cranes, are restricted to only one clutch per year and only rarely renest if their eggs or nestlings are lost to predators. Consequently, their annual reproductive potential is limited to one or two chicks per year. Females can be induced to lay more eggs if the eggs are removed from the nest before the clutch has been completed, which is how conservation workers persuade females to produce more young in captivity.

Incubation begins with the first egg laid, except in crowned cranes, which wait until the clutch is complete before they start. The incubation period in cranes is about 29 to 32 days (33 to 36 days in wattled cranes). In most species, both parents share incubation duties equally throughout the day, but one parent (usually the female) incubates primarily at night. Among whooping cranes, parents trade off about six to eight times a day, with incubation bouts lasting around two hours. An incubating crane sits upright with its head erect, surveying the horizon for danger. Occasionally, it will stand up and rearrange the eggs with its bill tip, which helps maintain uniform incubation temperature and prevents the embryo chick and its associated membranes from adhering to the inside of the eggshell.

The non-incubating parent also keeps watch and will defend the nest vigorously against potential predators, including feral cats and dogs, foxes, coyotes and humans, using a variety of approaches to thwart them. Sometimes, cranes "mob" intruders by descending rapidly on them while flailing their wings and uttering loud alarm calls, which is often a sufficient discouragement. A distraction display may be useful in luring predators away from the nest. Cranes may resort to a "broken wing" display for distraction, or they may dance the intruder away from their nest. From time to time, both parents participate, with one member of the pair mobbing the threat and the other attempting to lead it off simultaneously with a distraction display.

Development of the chick does not start until incubation begins. Consequently, crowned crane chicks hatch synchronously (at the same time) — even though the eggs were laid on subsequent days — because incubation, which occurs over a fixed length of time, began after the clutch was complete. Other species, however, hatch asynchronously (at different times) because incubation

began with the first egg laid. In this case, the chick that hatches first will be older and larger than its sibling.

When they are born, crane chicks are covered in soft, luxuriant down, which is tawny brown to cinnamon or silvery gray in color. They are strong and active shortly after hatching and are soon able to walk, run and swim. Both parents are devoted to their young throughout its prolonged period of growth and development. Almost as soon as the chick hatches, the parents are providing

Although downy sandhill crane chicks are able to run and swim soon after hatching, sometimes they prefer to rest amid the warmth and security of their parent's feathers.

it with food. Among its first meals are bits of eggshell, which the parents gently break apart and offer carefully to their youngster. Eggshells provide the chick with readily available calcium, which is not always easy to obtain from other sources. Calcium and protein are of utmost importance to developing chicks, particularly ones that must grow so rapidly.

Young cranes may leave the nest within hours of hatching, but often do not move out of the general area until they are a few days old. When the chick is sufficiently robust to travel, many crane parents lead their young — male in front, female second, chicks in tow — to the part of the territory where food is most plentiful. Parents feed chicks directly by offering them food bits, often mashed first so they are more palatable, with the tips of their bills. After a while, parents will encourage a chick to learn to feed itself by placing food on the ground and waiting patiently for the little bird to pick it up. Young chicks are also eager to watch their parents feed and frequently mimic their foraging behavior, although even older chicks are prone to begging incessantly for food when they have no inclination to search for it themselves. Nonetheless, crane parents remain gently

Newly hatched blue crane chicks are covered with fuzzy, grayish down that keeps them warm and provides some camouflage from predators.

devoted to their offspring and continue to feed them larger food items, such as fish, even when they are several months old and already fledged. Parents also brood (snuggle to keep warm) chicks at night and in poor weather. To stay warm, very young chicks often crawl up under feathers on their parent's back and sit with their head peeking out from under the wing's leading edge.

Although most cranes lay a clutch of at least two eggs, only one chick typically survives into autumn. Aggression between chicks is fierce and, unless food resources are unusually high, the dominant chick (generally the oldest one) outcompetes the younger for food. Some species are able to successfully raise two chicks in years of abundance by splitting them up so they have no contact with each other, each parent taking a chick and leading them to separate feeding sites. Many species successfully raise twins from about 10 percent to 30 percent of all clutches, although multiple young have not been observed in wattled cranes.

Fledging is the milestone marked by a young bird's first flight; the fledgling period is the number days required from hatching to flying. Most young cranes fledge at about two to four months of age, varying somewhat among species. Shortest fledgling periods, about 50 to 90 days, are found among cranes that nest either at high latitudes, including Siberian, whooping and lesser sandhill cranes, or high altitudes, such as demoiselle and blue cranes. These species have an abbreviated nesting cycle because poor weather descends early. Chicks must be full flighted and ready to move before autumn closes in. Wattled and sarus cranes and other tropical species have no need to be rushed and, consequently, have much longer fledgling periods (120 to 150 days). Juvenile cranes do not leave their parents' care when they achieve flight; on the contrary, they frequently stay with them throughout the winter. Young migratory cranes remain in their parents' company until they return to the breeding grounds in spring, after which they leave or are driven off the territory as the next nesting season begins. The young birds do not remain alone, however; juveniles of most species form loose nomadic flocks with other immature cranes where they spend the next two to four years, until they are old enough to seek a mate and establish a nesting territory.

Migration and Seasonal Movements

Migration provides birds with opportunities beyond the reach of most land-based animals, including plentiful food and moderate climate year-round. What is more, birds have the good fortune to thrive in what few people experience — the freedom to move at will, unimpeded by the constraints of local condition. Not all cranes are migratory, but those that are not still benefit from short periodic flights that are also intended to optimize feeding and nesting conditions. Many nonmigratory cranes, including blue, wattled, brolga, sarus and crowned cranes, move in flocks for short distances between preferred nesting habitats and nonbreeding feeding habitats. Often these seasonal movements are associated with oncoming rains that improve the abundance of local resources. Short seasonal flights are also characteristic of a few southern populations of more northerly species that are fully migratory. Such is the case with demoiselle cranes of northern Africa and sandhill cranes in Mississippi, Cuba and Florida.

Among the remaining species are archetypical long-distance migrants that heed their semiannual calling to travel some thousands of miles, often across inhospitable habitats, to better their environs. Many northern-breeding cranes are cold-hardy — one need only watch red-crowned cranes dancing in the snow in Hokkaido, Japan — but must migrate south as available food sources in the north diminish with winter's approach. The journeys of the long-distance migrants may be marvelous merely because they are arduous — some populations of pretty, diminutive demoiselle cranes migrate across Middle Eastern deserts and others navigate high passes of the Himalayas, the world's tallest mountain range. We can also wonder at the sheer length of their journeys — whooping cranes travel 2,500 miles (4,000 km) from the Canadian Northwest Territories to coastal Texas each autumn; Siberian cranes wintering in Iran migrate 3,400 miles (5,500 km) north across Eurasia to Siberia. As remarkable as these feats appear, they are diminished by the lesser sandhill crane, with the longest migration route among cranes, which carries the birds from marshy tundra nesting habitat in northeastern Siberia across the Bering Sea south to the playa lakes and riparian wetlands of northern Mexico, an annual round-trip of almost 14,000 miles (22,500 km).

Northern cranes, both Old World and New World species, adhere to a relatively predictable annual schedule, migrating south to wintering grounds in autumn and north again in spring to their breeding sites. Some species, including sandhill and Eurasian cranes, migrate in flocks of 50 to 100 individuals. Whooping cranes and others travel in much smaller numbers, typically family groups. Young birds depend on experienced adults to teach them the migratory route; thus, offspring fly in formation near their parents, constantly calling among themselves to remain coordinated and, perhaps, to receive much-needed reassurance. In subsequent years, after their parents have graduated older offspring to devote time to new chicks, juvenile cranes that are not yet old enough to seek mates often migrate in small flocks of older nonbreeding individuals, providing them ample opportunity to practice their newly acquired knowledge.

Although travel typically occurs in small groups, many cranes congregate before migration at traditional staging areas that have been used by species members for hundreds of generations. These places offer safe roosting spots and abundant food resources where, sometimes, thousands of birds may spend several days or weeks building up fat reserves and assimilating into flock society. Sandhill cranes are among the most remarkable of all staging species. Every March, for perhaps the past 10,000 years, almost a half a million sandhill cranes have met along a 75-mile (120 km) stretch of the North Platte River in southern Nebraska before they continue their long journey to their northern breeding grounds. Most birds will spend about four weeks gorging on waste grain in fields and meadows adjacent to the river. The hours near daybreak and just before sunset are filled with other important activities, including preening and dancing. At night the cranes roost about 4 feet (120 cm) apart, standing in shallow water along submerged sandbars or island edges where they are well protected from predators. Quality roosting spots may contain 30,000 birds at one time, and densities greater than 12,000 cranes per half-mile (0.8 km) of river commonly occur. Daily weather conditions strongly influence the start of migration; in particular, cranes like clear or partly cloudy days with gusty breezes from a favorable direction. Birds generally make use of high- or low-pressure weather systems that have a robust northerly or southerly flow to provide them with reliable tailwinds. Before leaving the staging area, cranes feed for several hours. They fly throughout the day, and then seek a secure overnight roosting spot in the late afternoon. The following morning, weather permitting, they continue their journey until they reach their final destination some hundreds or thousands of miles farther ahead. But this is just the beginning of the story, for migratory behavior in cranes is so intimately entwined in the fabric of these birds' life history that it cannot be so easily abridged. The true depth of this association, and the account of our enlightenment as humans, will be revealed in Chapter 4.

Demographics and Threats

Throughout the centuries, the largest birds are habitually exploited for their meat, the most beauteous for their feathers, and the most venerable for those parts of them that are perceived to confer what makes them worthy of our reverence. Cranes have met their demise for all of these reasons. Traditional hunters in Europe used trained falcons or small eagles to force them to the ground so that dogs could dispatch them; gyrfalcons of North America (gyr = crane) may have been named after this practice. Millions of feathers were removed from whooping cranes and other species, killed while roosting, to be shipped to American and European cities, where they frivolously adorned womens' hats. Cranes were cursed with flesh and fiber thought to have curative properties. Their meat was considered effective relief from cancer, ulcers, palsies and "winde in the guts." Bone marrow was used as eye salve, and their fat was placed in ears to cure deafness. Their "gall" was considered most useful in treating all nature of maladies, including forgetfulness. Moreover, cranes fell prey to the most foolish of practices — the belief that consuming part of an animal will somehow confer its finest characteristics. For this result, cranes were killed for "nerves" in their wings, which were thought to make one strong, and their eggs were destroyed to prepare an elixir of immortality.

Yet none of these activities compares to the devastation now facing crane populations. Cranes are among the most severely imperiled of all bird families. Six of 15 species are considered vulnerable by the World Conservation Union, formerly the International Union for the Conservation of Nature (IUCN), in all or parts of their range; three more species are endangered, with global populations numbering less then 3,000 individuals. All species are declining precipitously globally or locally, save one: the whooping crane, which is clawing back from the brink of extinction due to extensive conservation intervention measures.

The anthropogenic threats against crane populations are pervasive and multifaceted, and there are differences between species. Nonetheless, the greatest injury is the result of destruction, degradation and modification of prime nesting and wintering habitat — particularly wetlands and grasslands — in conjunction with direct losses through hunting, pollution, poisoning and encroachment. Cranes that occur near large human populations, such as in Southeast Asia, are at greatest risk.

Upstream water development projects on the Zambezi River have so modified downstream wetland habitats that nesting wattled cranes often have insufficient food for their young.

Virtually all cranes rely on wetland habitat for resources required for survival, and several species are utterly dependent on it. However, wetlands are universally underestimated in their importance to wildlife — many people see them just as useless "swamps" — and, consequently, they are being destroyed or degraded at alarming rates around the world. Wetlands are frequently lost through conversion for other uses including agriculture, urbanization and other commercial development and recreational activities; the conversion process typically involves water drainage and removal of vegetation. Drainage of wetlands, particularly for agricultural purposes, is harmful for most crane species as it eliminates their use for feeding, nesting and roosting. Partial conversion can also be detrimental because fencing on fields and pastures adjacent to remaining wetlands can ensnare flightless young or separate them from their parents, ultimately leading to starvation and death. Brolgas and red-crowned cranes are most seriously affected in this way. Species that are able to subsist on grain and other crops suffer less when wetlands are converted to agriculture; however, these birds frequently perish by the gun or through intentional poisoning if farmers consider them nuisances.

Dams and water diversion projects degrade wetland habitats by severely altering stream-channel characteristics. Damage to habitat upstream of a dam when it is flooded to create a reservoir is well understood; habitat changes downstream are less easy to predict. Often the controlled reduction of water flow below the dam causes irreparable damage to wetlands, riparian (river bank) and lacustrine (lake) ecosystems because it eliminates seasonal cycles of flooding and sedimentation that maintain the integrity of these habitats. Dam construction and water diversion have already reduced viable crane habitat along many of the world's largest freshwater systems, including the Danube, Zambezi, Senegal and Platte rivers and Lake Chad. Other rivers that additionally harbor crucial crane habitat, such as the Mekong, Yangtze and Amur rivers, the Okavango system and the Sudd wetlands, are currently under threat of development. Moreover, these systems often provide critical habitat for more than one crane species. Five species, including four endangered or vulnerable species, winter at Lake Poyang and on the Yangtze River delta in China. These habitats are maintained by annual flooding and by sediments moving down the river that create several hundred square feet of land every year to overcome the loss through erosion. The Three Gorges Dam, the world's largest, is currently under construction upstream from these prime wintering habitats. When completed, the annual net loss of freshwater wetlands, mudflats and saltwater marshes will steadily deprive many Eurasian and hooded cranes, large populations of red-crowned and white-naped cranes, and virtually the entire world population of Siberian cranes of wintering habitat. Precipitous declines in these species have been forecast as a result. Plans are currently underway to build a dam larger than the Three Gorges Dam along the Amur River on the China–Russia border. Nearly half of all crane species spend some time along the

Amur, including many that winter on the Yangtze River. Two such projects within the geographic distribution of any wildlife species would hardly be sustainable. The destructive effect of river damming is not restricted to Asia. On the contrary, upstream dams on Nebraska's Platte River have so moderated the flood cycle that unsuitable dense woody vegetation is growing where open riparian habitat used to exist. This has substantially reduced the availability of preferred roosting sites for endangered whooping cranes and hundreds of thousands of migrating sandhill cranes.

Unfortunately, large birds such as cranes are perpetually under pressure from hunting because an individual kill provides a sizable portion of meat. Subsistence hunting of cranes occurs globally, including in Afghanistan, Canada, Nepal, Russia and many African countries, and increases in importance with war and political unrest and poverty. Poaching is also widespread because laws set in place to protect cranes are typically ignored and rarely enforced. Furthermore, black market trade in crane parts and eggs for preparing medicinal remedies, and crane feathers for making fans in Hong Kong, are taking their toll on many crane species.

In light of the cranes' global plight, it is surprising that sport hunting is legal in many countries, although it is not without major controversy. Among the species that can be legally hunted are sandhill cranes in Canada, the United States and Mexico, and Eurasian and demoiselle cranes in Pakistan. It is generally considered that species abundance is an adequate hedge against the direct loss of birds to hunters. However, hunting can place undue stress on local populations even if the species, as a whole, appears plentiful. Furthermore, lack of accurate information regarding total species and population numbers and overall kill rates can lead to underestimating the effect of hunting on species survival. For example, some crane populations have been extirpated (locally driven to extinction) by sport hunting. During the 1990s, hunters in Afghanistan and Pakistan decimated the Siberian crane population that wintered in India as they migrated through each spring and autumn. Hunting has been a popular sport among the wealthy of Pakistan for centuries; however, the traditional hunting practice of throwing a *soya*, rope with weighted ends, required some skill to kill a crane, thus limiting the number of birds taken. Now an increasing number of hunters, particularly in the tribal areas of the North-West Frontier Province of Pakistan, are using guns to kill cranes. Undoubtedly, more cranes are falling than before.

En-route to their wintering grounds in the southern United States and Mexico, three subspecies of migratory sandhill cranes, totaling about 450,000 birds, grace the skies of central North America during autumn hunting season. Tens of thousands of cranes, many juveniles, never arrive. Although some wildlife managers believe that this is a sustainable kill, most of the dead birds are greater sandhill cranes, a subspecies comprising about 250,000 individuals. This is due to differential timing of migration between the three subspecies. Greater sandhill cranes migrate in September; however, Canadian and lesser sandhill cranes pass through in October, after the hunting season has closed. Whooping cranes also migrate in October. The crane-hunting season was shortened after it was discovered that an untold number of endangered whooping cranes were killed every year — and concealed through fear of prosecution — by hunters who were unable to correctly distinguish a large, brilliantly white crane from a much smaller gray one. Precious whooping cranes still fall to some hunters who mistake a 5-foot (160 cm) tall, long-legged bird with a 7-foot (210 cm) wingspan for a goose. Herein lies another degree of inaccuracy in the annual crane hunt. Unofficial reports suggest that unretrieved birds, which are crippled by hunters and left to die, may number at least 15 percent of the retrieved kill, which are those birds comprising a hunter's legal quota. Hence, sandhill cranes that are taken by the gun each autumn may approach 35,000 birds. One would question the sustainability of these numbers in our rapidly deteriorating environment.

Aspects of crane ecology — in particular their reliance on wetland and agricultural habitats for food — increase their exposure and susceptibility to poisons and other pollutants. In addition, the bioaccumulation, or increasing bodily concentrations, of these toxic substances can kill cranes directly, weaken them by impairing physiological processes and reduce their reproductive capacity. Accidental poisoning through ingestion and exposure to pesticides in agricultural fields occurs frequently when they eat tainted seeds, grains and insects. Furthermore, many cranes, principally African species — grey crowned, blue, demoiselle and wattled cranes — are killed in great numbers by farmers through intentional poisoning to prevent crop damage. Detrimental affects due to chemical and organic environmental toxins from household, agricultural and industrial pollution are of long-term concern because they degrade prime wetland habitat by affecting water quality and tainting aquatic invertebrates and fish that are eaten. Catastrophic spills of toxic substances are acute and unpredictable problems. Unfortunately, chemical processing and transportation facilities frequently occur along waterways adjacent to crane habitat. Of particular concern is the Gulf Intracoastal Waterway in Texas. Barges carrying highly toxic substances, such as benzene and xylene, pass daily along the waterway adjacent to the Aransas National Wildlife Refuge, the winter home of all wild whooping cranes. An accidental spill would be irreparable to this fragile species and its habitat.

A myriad of other human actions threaten cranes worldwide. For example, the traditional practice of keeping cranes in captivity as pets or for pest control or food, which dates back to ancient Greek, Egyptian and Chinese civilizations, still occurs with great frequency, continually fueling the burgeoning illegal live bird trade. Blue, wattled, crowned, demoiselle, sarus and brolga cranes, typically prefledged juveniles, are most frequently trapped for this purpose. Other, more obliquely human-caused effects are no less damaging. These include collisions with power lines and other structures that maim or kill migrants; human disturbance that causes young

to be lost to predators when adults are flushed continually from the nest; war and other political upheaval that increase direct mortality through subsistence hunting and indirectly through disturbance and habitat degradation; and increased predation though both intentional and accidental introduction of exotic animals, such as mink and other members of the weasel family. By no means is this a complete list; the anthropogenic factors affecting cranes and other wildlife globally are both pervasive and insidious. One could certainly paint a dismal, irrevocable picture for the future of cranes if it were not for efforts of a few handfuls of biodiversity conservation workers, politicians and private citizens around the world who are dedicated to saving cranes. Perhaps there is still hope.

Crane Conservation

Cranes embody all nature of challenge in biodiversity conservation. They are migratory birds that span many political boundaries on their semiannual journeys, they conflict with humans for resources and they are subject to exploitation by the black market trade. In addition, they suffer from multifaceted threats and habitat loss and degradation on all levels of the ecosystem: in local, regional and global horizons. Furthermore, many species already occur in small populations that are barely viable genetically, and the basic life history strategies of cranes belie the efforts of workers to increase and stabilize them. Yet there are factors working in their benefit. Cranes are highly conspicuous birds, and ornithologists have already amassed considerable scientific knowledge for most species. Moreover, the magical beauty of cranes is universally compelling, and any conservation concerns can easily attract attention, if appropriately disclosed. Cranes are, thus, emblematic flagship species for several conservation organizations.

Three great white cranes represent the world's most imperiled species. The whooping crane is the most endangered of these, numbering about 335 wild birds. Its sister species, the endangered red-crowned crane, is the second rarest, with populations totaling about 2,000. Nonetheless, the critically endangered Siberian crane is considered the most seriously threatened, not only due to its rarity (about 3,000 birds) but also because it is entirely dependent on wetland habitats that are on the brink of destruction. These, and other vulnerable cranes, have been the focus of concerted conservation efforts in recent decades.

Many crane conservation programs target increasing population size. Small populations face sudden detriment loss due to stochastic, or random, events including disease outbreaks and natural and human-made disasters. Those that occur in a restricted area, such as endangered Mississippi and Cuban sandhill cranes, are at greatest risk. In addition, small populations exhibit increased incidence of genetic disease or malformation, depressed immune function and

Opposite: West African black crowned cranes are among the three crane species that cannot be bred reliably in captivity, causing additional concerns when declining populations require rigorous conservation intervention.

infertility due to inbreeding (breeding among closely related individuals) and often have unnaturally skewed gender or age-class ratios. Unfortunately, desperately small crane populations have a tendency to stay small or disappear entirely without intervention. Cranes have among the lowest avian recruitment rate; in other words, new birds are added by birth only very slowly. In a typical crane population, nonbreeding juveniles average about 10 to 20 percent; breeding usually does not occur until about five years of age. Each pair of adults can raise about one chick per year, so, essentially, only about one-quarter of the breeding population is producing a chick at any time. If losses due to hunting, disease, inclement weather and migration are considered, then the ability of crane populations to increase their numbers substantially is extremely limited.

For this reason, captive breeding programs are maintained. All crane species except West African black crowned cranes and wattled and hooded cranes can be bred reliably in captivity. Captive flocks serve as both a reservoir for genetic variation in critically imperiled species and a hedge against total annihilation. In addition, birds raised in captivity often can be released into the wild to bolster small wild populations. Initially, eggs were taken from wild nests to generate many captive flocks. Recent propagation successes have established sufficient captive breeders to produce eggs for release programs, and the collection of wild eggs is a now a relatively rare event.

Captive-bred cranes of six species are currently being placed into the wild in numerous reintroduction programs. Active releases are being considered for at least three more species. Reintroduction of non-migratory species is generally quite successful, although increased predation on naïve newly released cranes is worrisome. However, the difficulty arises among migratory species whose young naturally rely on experienced wild adults to teach them the traditional route passed down through generations for millennia. Migration is part of "wildness" in birds and, although sedentary populations of previously migratory species are better than none at all, they are really little more than zoo specimens living in a natural environment. It was critical that captive breeding programs determine how to teach birds to migrate. Cross-fostering endangered species using wild migratory parents of a common species has been attempted, but results were disappointing. Cross-fostered young generally survived and many migrated, but when choosing mates they selected individuals from the surrogate species that raised them, not from their own. It became clear that captive-bred cranes would have to be shown how and where to migrate. So began many experiments using guide-birds, hang gliders, all-terrain vehicles, trucks and ultralight aircraft to lead cranes to their winter homes. Only one method has been clearly successful, which will be discussed in considerable detail in Chapter 4.

Loss and degradation of habitat is the primary cause of decline in most birds. One reason this occurs is because, as preferred breeding habitat is lost, the number of acceptable nesting territories also declines. Many species, including cranes, will not breed unless they can acquire an acceptable territory that meets the needs of them and their young. Fewer territories mean fewer young recruited into the population. Furthermore, annual breeding success is highly variable, being dependent somewhat on the territory quality and the resources that it offers. Degraded breeding territories also produce lower juvenile recruitment. During migration and in the nonbreeding season, preferred habitat in adequate quantity and quality is required for foraging and roosting; otherwise, there is decreased survival due to starvation and predation. Consequently, conservation programs must focus on preservation of prime habitat required throughout the year to maintain a viable core population. And too, since cranes utilize considerable space, efforts to protect and restore function to critical ecosystems must be directed to a larger landscape, perhaps in the order of entire watersheds. The complexity and difficulty of this task is augmented in migratory species that span geographic and political boundaries.

For these reasons, crane conservation has often required international cooperation and the formation of task groups to implement challenging conservation measures that are required to see cranes through the 21st century: habitat preservation and management; legal and cultural protection; international agreements; and surveys, census and global research. One such organization is the International Crane Foundation (ICF), which was established in Baraboo, Wisconsin, in 1973 by Dr. George Archibald and his colleague the late Dr. Ron Sauey. The ICF is committed to crane conservation through research, education and protection of wetland and grassland habitats on which the birds depend. It represents the World Conservation Union/ Species Survival Commission Crane Specialist Group, which currently has 79 volunteer members — experts on all aspects of crane biology — from 28 nations that possess not only knowledge but also a deep commitment to saving cranes.

The conservation of imperiled wildlife presents bittersweet challenges; the struggles are great and the tangible successes are few. Yet the long road back for many species offers benefits beyond a simple increase in population numbers, and such is the case for cranes. What we learn along the way through research, education and communication is an invaluable fortune to carry into the future. In addition, the protection of habitat profits not only target species but all other animals, plants and ecological processes that share their realm. Even a small measure of international cooperation yields enlightenment and tolerance in a world ravaged by political and cultural conflict. But cranes themselves add the last piece to the puzzle. As flagship species, their grace and beauty personify the value of the endeavor. Let us hope that cranes may inspire global environmental awareness, and that they may be the ambassadors of peace to unite the world's accountability to wildlife conservation.

Chapter 2

The Decline of the Whooping Crane

Those of us fortunate to see whooping cranes in life think of them as marsh birds, for this is where our mind's picture will place them — head down, walking deliberately among the reeds, a stately profile set upon a backdrop of green and gold. It is true that they are somewhat more reliant on wetland habitats than many other crane species, but, on a continental scale, whooping cranes are birds of North American grasslands. Their prairie biome formed through geological time where rainfall was low or occurred sporadically, and soil conditions favored grasses over other types of vegetation. Periodic fire was an integral part of grassland ecology because it suppressed the invasion of trees and shrubs. On the eastern tallgrass prairies, big bluestem grass fed the bison; its slender, blue-green stems grew 10 feet (3 m) high with red-tipped blades that turned bronze in autumn. It was among these vast oceans of wind-blown grass that bald eagles, prairie wolves and badgers lived in abundance. Rainfall declined to the west, as did the size of grasses and their diversity. Mixed-grass and shortgrass prairie formed there, where antelopes, swift foxes and burrowing owls trod on sun-baked soils and endured the aridity of blistering summers. It is no mere coincidence that the whooping crane's fate parallels that of its native habitat. Grasslands are the most endangered biome in North America; over 90 percent of this inimitable habitat has been converted to agriculture since 1837, when Illinois blacksmith John Deere invented a steel-bladed plow capable of tilling its rich, heavy topsoil.

Yet when they both prospered, the cranes and the prairies, whoopers sought out the most precious commodity that dry grasslands could offer: prairie wetlands — potholes, marshes and sloughs — that formed as still water accumulated in landscape depressions lingering after the retreat of Wisconsin glaciers some 10,000 years ago. Historically, whooping cranes found their nesting sites among millions of isolated water bodies that were ringed with lush vegetation. Their principal breeding range stretched across North America from what is now central Alberta through southern Saskatchewan and Manitoba, south through northeastern North Dakota, western Minnesota, southern Wisconsin, northern Iowa and northern Illinois. Whooping cranes also nested in the remote Canadian wilderness south of Great Slave Lake in the Northwest Territories and northern Alberta. In these northerly reaches, cranes laid claim to aspen parkland, a transitional habitat between prairie and boreal forest where

The sandbars and mudflats of Nebraska's Platte River provide safe roosts and plentiful food for thousands of migratory waterfowl and waders, such as whooping and sandhill cranes.

aspen, poplar and spruce groves dotting the barren canvas struggle for coexistence with hardy grassland species.

Although northern breeding grounds may offer mated pairs of whooping cranes safe haven to rear their young and ample food on which they may thrive, their summers are fleeting. Thus, as luminous white feathers begin to emerge on the wings, back and belly of their rusty young, the cranes turn their eyes to the south. Nearly all whooping cranes are migratory, and each autumn they join many billions of birds that travel the continent to seek more equitable winter environs. Departure from the breeding grounds occurs in September or October, with family groups lingering somewhat longer to ensure the hardiness of their fledglings before beginning the arduous journey. Whooping cranes do not migrate nonstop as do some smaller birds, and, unlike sandhill cranes, they rarely travel in groups larger than a few birds, although flocks of 10 to 40 individuals have been reported historically. The trip south may take two to seven weeks all told, with some legs of the passage being covered more quickly than others. Poor weather may slow the birds' southward progress or halt it for any number of days until clear weather returns.

When Robert Allen, biologist and ardent conservationist, described migrating whooping cranes as an experience packed with beauty and drama, he captured the thoughts of observers through the centuries who have witnessed these birds in flight: their startling white plumage, the depth of their wingbeats, their shrill, bugling *ker-loo! ker-lee-loo!* call. Knowledge of the seasonal movements of less conspicuous birds may be gone forever as the species themselves are lost; however, this is not the fate of cranes. Whooping cranes have been frequently seen in traditional staging areas along their route, such as in Saskatchewan and Nebraska, where they remain for several days to rest and feed. The importance of the Platte River valley in Nebraska to migrating cranes has been known for over 200 years. The river's wide channel offers roosting and loafing birds the relative safety of low, exposed sandbars surrounded by shallow water, and there are nearby wetlands and croplands with good visibility and ample food where cranes may forage. It is tragic, however, that this traditional knowledge of whooping crane behavior is used so frequently to their detriment. Nebraska scores an untold number of whooping cranes shot by hunters, second only to Louisiana and Texas, where the birds have spent their winters.

Nevertheless, from many scattered, often imprecise, reports — taken for good or ill — we can reconstruct at least two historical migration routes that led whooping cranes to two primary wintering sites. Cranes that nested in the Midwestern states likely wintered in the tallgrass prairie and salt marshes of southwestern Louisiana. Birds bound for coastal inlets and beaches on the Gulf Coast of Texas had journeyed from northern United States and Canada. More disparate reports imply there may have been other wintering grounds in the northern Mexican highlands, and on the Atlantic Coast in New Jersey, Delaware, South Carolina and Georgia, where they have not been seen for a century or more.

Migrating whooping cranes complete their journey from mid-October to early November; family groups usually arrive on the wintering grounds two or more weeks later when traveling with their young of the year. Each mated pair or family group quickly establishes a winter territory of several hundred acres where they feed primarily on blue crabs, clams and shrimp among the coastal lagoons and beaches, freshwater or brackish marshes, or tallgrass prairie of their winter homes. With luck, winter passes uneventfully and spring once again brings on migratory restlessness. Mated pairs perform their leaping dance and call to each other in unison to renew their perennial pair bond; young birds that are not yet ready to breed are impatient with anticipation. Family groups are the first to depart in late March, with others following through April. Thus, spring migration proceeds rapidly. Experienced cranes may cover the distance they traveled in several weeks in autumn in merely 10 days in spring. The distance covered by whooping cranes from summer to winter and winter to summer measures at least 48 degrees of latitude, some 6,600 miles (9,600 km) return, yet it was not the journey's length that spelled the fate of whooping cranes, but the perils that they faced along the way.

Whooping cranes have been gracing our skies since time immemorial, with a fossil record that dates back to the Upper Pliocene of southern Idaho, some 3.5 million years ago. The Pliocene world was warmer than today's, perhaps due to higher ocean surface temperatures or greater concentration of carbon dioxide and other greenhouse gases in the atmosphere. However, as the epoch progressed, the global climate gradually cooled and ice caps formed at both poles. This exacerbated the cooling trend as more of the sun's radiant heat was reflected off snow and ice. The earth's biomes changed considerably; most importantly for crane evolution, the cooler, drier seasonal global climate favored the proliferation of grassland and savanna habitats on all continents (except Antarctica) where tropical vegetation once thrived. The Pliocene crane was virtually indistinguishable from modern whoopers. Even then, this crane fed and reared its young in the lush marshes that ringed quiet shores of an ancient lake. It shared this pristine world with beaver, otter, terrapin, geese and swans, pelicans and cormorants, and other aquatic species. Yet the peace would remain for many millennia, because as this crane searched for crawfish among cattails and bulrushes, we humans were just learning to walk upright.

The cooling trend that began during the Pliocene continued into the Pleistocene epoch, which occurred about 1.8 million to 12,000 years ago. The Pleistocene was typified by repeated glacial cycles, or ice ages, which covered up to one-third of the earth's surface with continental ice sheets that were 5,000 to 10,000 feet (1,500 to 3,000 m) thick. As the ice sheets retreated during interglacial periods, they left behind a blank canvas studded with chunks of stranded ice. Opportunistic grassland and savanna vegetation

The creation legend of the Muscogee (Creek) Nation features a white crane that is honored for both its wisdom and illustriousness.

colonized readily in the absence of forest, and as the remaining ice melted, it formed small, water-filled depressions called potholes.

Whooping cranes thrived in this landscape of broad savannas and wet prairies. Their Pleistocene range overall was larger than in recent history, their fossils sketching a distribution that stretched from southern California north to Idaho and east to Michigan, Kentucky and Florida, and included representatives from Arizona, North Dakota, Kansas, Illinois and Virginia. Fossils from the Florida Panhandle east and south to the central Atlantic Coast suggest that this region offered bountiful freshwater ponds and marsh habitats during the Pleistocene. Others found near West Palm Beach date from 21,000 years ago and are among the most southerly Pleistocene avifauna fossils in the United States. These cranes lived during the last glacial period, the Wisconsin, when the continental ice sheets had pushed south to their maximum extent. They shared the moderate Florida climate, some 1,000 miles (1,600 km) from the glacier's edge, with a myriad of birds and beasts now extinct or well beyond their range, such as giant mammoths, ancient bison, tapirs and condors. Historically, whooping cranes may never have been as abundant as their smaller cousins the sandhill cranes. In some Pleistocene deposits where both species coexisted, sandhills outnumbered whooping cranes 29 to 1. However if this ratio somehow reflected a perfect world, there would be 17,000 whooping cranes alive today, which, of course, there are not.

Whooping cranes had been dancing the circle of their lives for millions of years before humans crossed the Beringia Land Bridge that connected Asia and North America only 13,000 years ago. Yet the first people to live among cranes trod lightly on their fate. But they knew the white cranes well; they were *tut-teeghuk* to the Inuit of Hudson Bay and *wab-a-gi-gak* to the Chippewa peoples of Minnesota. Their name was *wapo oocheechawk* in the Cree language, *tarakowa* in the voice of Algonquin and *Peto cka* to the Osage. The peoples of the First Nations bestowed great wisdom on whooping cranes and they were often revered as healers. The creation legend of the Muscogee Nation tells how all other animals vied for the First Child's attentions by anxiously adorning themselves with color. The white crane, however, kept fishing, and thus was honored with the role of healer, befitting one of such great wisdom.

Traditional aboriginal folklore often featured whooping cranes in tales that served to explain some remarkable aspect of life history while teaching a critical lesson. "The Great Ball Game" of the Muscogee Nation illustrates the ills of boastfulness, an offense for which Crane was forced to migrate south each autumn. Bragging is also punished in the Sioux story of the frogs and the crane, when the frog that declares loudly that he is the chief of the pond comes under the watchful eyes of a hungry crane. Finally, Cherokee legends tell of a race between the crane and the hummingbird that is not unlike Aesop's story of the tortoise and the hare — Hummingbird is swift, but Crane is steady because he flies through the night when Hummingbird must stop. The prize for finishing first is the affections of a beautiful woman. Although Crane wins the race, both birds lose because the woman decides not to marry after all when she realizes that the victor was awkward Crane, not handsome Hummingbird.

There is evidence among prehistoric archeological sites in Illinois, North Dakota, Georgia and Kentucky that whooping cranes were occasionally hunted or their eggs eaten. However, the rarity of these finds suggests that First Nations peoples did not substantially affect whooping crane populations. Eating the birds was taboo among some aboriginal peoples. In his 1819–20 expedition from Pittsburgh to the Rocky Mountains, explorer Edwin James encountered Omawhaw Nation peoples. His journals describe their culture, in which one clan, the Hun-guh, took whooping cranes as their spiritual symbol. They were strictly forbidden to eat them, an offense punishable by blindness, gray hairs and general misfortune. Even mistakenly tasting or eating this food would sicken the consumer as well as his family. It is not among the First Nations peoples that we find the devastation of the white crane.

There are many billions of humans on earth that are fully able to deny the undeniable fact that we, as a species, are almost entirely responsible for the ills of our world, by never affixing blame where it is deserved. Many people believe that these things just happen during the day-to-day existence of our planet, that somehow something else did something wrong: "the bird was too set in its ways and was unable to deal with environmental change; the species was too specialized for its own good; they were never very common anyway and it was already going extinct before we arrived." But the fact remains that whooping cranes thrived in North America before the first European explorers and colonists set foot on the Atlantic shores; the whoopers did not change — we did.

[The whooping crane], unchanged, moved through these shifting scenes with the same nobility, the same dignity we know today. After unknown generations of existence, the drama — the tragic drama — of its meeting with 'civilized' man — was about to unfold.
— Robert Allen —

The journals and captains' logs of the first Europeans to land in the New World contain a great many references to white cranes. Over 400 years ago, these visitors beheld them, first in North Carolina, then later in Ontario, New Jersey, South Carolina, Georgia, Florida and the Northwest Territories. Reports of these early encounters are often misleading; eastern North America is home to several other species of primarily white, long-legged birds, including the white ibis (*Eudocimus albus*), wood stork (*Mycteria americana*) and snowy (*Egretta thula*), cattle (*Bubulcus ibis*) and great (*Ardea alba*) egrets. Nonetheless, as historians, we are fortunate that the sheer splendor of whooping cranes made them noteworthy, as this affords some opportunity to assemble the phantom that is this bird's past. Unfortunately, it was this magnificence that nearly cinched its destruction.

In mid-July in 1584, Captain Philip Amadas made landfall on Wokokon Island (now know as Ocracoke Island) in the Outer Banks, the barrier islands off the North Carolina coast. He had left England three months before in a party of two tall-masted barques; Amadas was captain of the ship *Tiger*. Captain Arthur Barlowe skippered the other ship, the *Admiral*. Both men were under the command of Sir Walter Raleigh, who had been granted a royal patent in March of 1584 to colonize new lands in the name of Queen Elizabeth I.

As they came ashore, the men admired the island's pristine beaches; its marshes and tidal flats were teeming with waterfowl and wading birds. The captain's log entry reads:

> *Under the banke or hill wheron we stode, we behelde the vallyes replenished with goodly Cedar trees, and having discharged our harquebuz-shot, such a flocke of Cranes (the most part white) arose under us, with such a cry redoubled by many ecchoes, as if an armie of men had showted all together.*

Amadas and his men fired but, fortunately, their primitive weapons were unable to secure a kill. Thus, there remains no lasting evidence to confirm this first encounter, save the logbook entry, which suggests that whooping cranes bred in North Carolina some 400 years ago.

The first European encounter with whooping cranes in Canada occurred about 30 years later in 1615 as Samuel de Champlain explored the Great Lake of the Entouhonorons (Lake Ontario). On at least two occasions in autumn, Champlain and his men sighted "many cranes which were as white as swans" at the lake's eastern end and 12 leagues (about 36 miles or 60 km) upstream on the Cataraqui River, perhaps near the Rideau Lakes. Although details are few, one would wonder if these were migrant flocks staging in the little lakes and isolated marshes that are plentiful in this region of Canada. Scattered reports of whooping cranes in Ontario from 1871 to 1900 lend some credence to Champlain's journal entries.

But as Samuel de Champlain sailed the lower Great Lakes, other European explorers navigated the treacherous, ice-laden Arctic Ocean seeking the Northwest Passage. Since the 15th century, Europeans had been searching the archipelagos between Greenland and Canada for an alternate trade route across the North American continent to Asia, which would shorten the existing journey around Cape Horn, at the southern tip of South America, by many thousands of miles. In July of 1631, Captain Luke Fox, an English explorer, was sailing the western shores of Hudson Bay when his crew brought aboard a strange bird. His journal describes the event:

> *They brought on board two goodly Swannes, and a young Tall Fowle alive, it was long headed, long neckt, and a body almost unanswerable; for it was but pinfeathered. I could not discerne whether it was an Estridge or no, within 3 or 4 dayes the legges by mischance were broken, and it dyed.*

Fox never determined the nature of his "ostrich" nor did he find the Northwest Passage. Shortly after this encounter, he decided that no such route to Asia was possible. He was wrong on both counts; his bird was probably a juvenile whooping crane and the Northwest Passage was finally breached in 1905 by Norwegian explorer Roald Amundsen. Unfortunately, these sightings of whooping cranes well beyond their current range remain tantalizingly out of reach because no specimens were brought back to Europe, despite many birds perishing at the hands of hungry seamen. This significant event would require the prudence of a naturalist.

Mark Catesby was the first true naturalist in North America. He had originally traveled from his native England in 1712 to visit his sister in Virginia. However, Catesby became so intrigued by the strange flora and fauna of the New World that he returned to America in 1722 to study it in more detail, having first received funding from benefactors such as Sir Hans Sloane, who later founded the British Museum. Catesby arrived in Charleston, South Carolina, in May and traveled inland, meticulously describing and illustrating animals and plants that he encountered. Unlike famous artist and ornithologist John James Audubon, who customarily killed birds and took them home to paint them with life-like poses and backgrounds, Catesby worked in the field, drawing all of his subjects from life — with one significant exception.

On his travels, Catesby was given the entire skin of a large, white un-described bird by an aboriginal man who had used it as his tobacco pouch. He told Catesby that great multitudes frequented the lower river in spring. This description was later confirmed by another man who had seen white birds farther south at the mouths of the Savannah, Aratemaha (Altamaha) and other rivers near St. Augustine, Florida; he added that they make a "re-markable hooping noise." Catesby made detailed drawings of the bird's head and neck, but he never saw another. Recognizing that it was a new species, he christened the bird *Grus americana alba* — the American white crane. A full description of the bird appeared in Catesby's beautifully illustrated book *The Natural History of Carolina, Florida and the Bahama Islands*. The volumes were first published between 1731 and 1743 and contained 220 prints depicting 109 different species. So admired was this work that Carolus Linnaeus — the renowned Swedish botanist known as the father of modern taxonomy — used 75 of Catesby's accounts on which to base his system of scientific binomial nomenclature for American species. However, not everyone was sufficiently academic to embrace Latin names, so on page 75 of Volume I, beneath the scientific species name, Catesby provided his readers with an English common name for this great bird. He called it the Hooping Crane.

An entire "hooping crane" was finally illustrated in 1743 in *A Natural History of Uncommon Birds*; the author, George Edwards, was a British naturalist who had recently gained some reputation for his painting. Edwards was given a well-preserved dried whooping crane skin by a Mr. Isham, who had acquired the bird near Hudson's Bay, where he claimed they could be found in summer. Edwards concluded that these cranes were, indeed, birds of passage that bred in the north and traveled south as winter approached.

Exploration in the Canadian north continued throughout the 1700s as the fever to find the Northwest Passage increased. The voyagers often encountered the curious "hooping cranes" in spring and summer; unfortunately, many meetings ended in the stew pot. As a French explorer so succinctly explained in 1744, "We have cranes of two colors [in Canada]; some are all white, the others pale grey, all make excellent soup."

Some men, however, had the forbearance to observe before they shot; one was English explorer Samuel Hearne. Hearne first saw whooping cranes while in the employ of the Hudson's Bay Company. He had been charged with finding bountiful copper mines purportedly known to local aboriginal peoples; thus, he became the first European to travel deep into the Great Slave Lake region of Canada's Northwest Territories. Although Hearne did not consider himself a naturalist, he was intensely interested in the natural world, and kept remarkably accurate journals of his observations. Hearne had watched whooping cranes near Hudson Bay in the spring of 1770 and 1771. He noted that they were generally seen only in pairs as they frequented open swamps, riverbanks and edges of lakes and ponds, searching for small fish and frogs. He added that they seldom had more than two young, and that they departed for the south early in autumn. Hearne may not be remembered for his contribution to the Canadian copper industry — his greatly sought-after mine yielded only 5 pounds (2 kg) of ore — however, he may be recalled as the first person to describe a whooping crane from life, not from the cadaver of a museum specimen.

As the new century dawned, explorers turned their eyes to the American west. The Louisiana Purchase of 1803 had secured for the Americans over 800,000 square miles (2 million square km) of uncharted territory west of Illinois to the Rocky Mountains; France received 3 cents per acre (7 cents per hectare) in the transaction. A few weeks after the deal was struck, President Thomas Jefferson appropriated $2,500 to fund the exploration of the territory, particularly the Missouri River watershed, to determine if there was a navigable water route to the Pacific Ocean for trade purposes. He selected Captain Meriwether Lewis and Second Lieutenant William Clark of the US Army to lead the expedition. Guided by the Shoshone Nation woman Sacagawea, Lewis and Clark traveled 8,000 miles (12,800 km) in 28 months from 1804 to 1806 across the western wilderness. On their journey, they described 178 plant and 122 animal species and subspecies — one of which was the whooping crane.

The cranes were first sighted on the upper Missouri River in the spring of 1805. Lewis' April 11 journal entry read:

Sacagawea of the Shoshone Nation guided explorers Lewis and Clark through the early-19th-century American west, where they observed many novel plant and animal species, including the whooping crane.

... saw some large white cranes pass up the river. These are the largest birds of that genus common to the country through which the Missouri and Mississippi pass. They are perfectly white except the large feathers of the first two joints of the wing, which are black.

No doubt, Lewis and Clark observed whooping cranes that day.

Explorers were not people of science, and, consequently, many of their descriptions were vague or inaccurate. We may consider these accounts when we are reconstructing a species' distribution or life history, but we cannot rely on them. Early in the 19th century, however, there dawned a new science: ornithology — the study of birds. Although ornithology at this time was a discipline based on observation, not scientific experimentation as it is today, it still fostered what we consider the "expert," someone who has devoted considerable study and contemplation to a particular subject. Occasionally even experts are incorrect; unfortunately, when this happens it may influence a generation.

Thomas Nuttall was an English botanist turned ornithologist who settled in the United States in 1808. During the next decades, he traveled extensively throughout the country, including the Hawaiian Islands, collecting plant specimens for the Liverpool Botanical Gardens and making notes of his additional observations. In 1811, he wrote of his first encounter with cranes:

> In the month of December, 1811, while leisurely descending on the bosom of the Mississippi ... I had the opportunity of witnessing one of these vast migrations of the Whooping Cranes, assembled by many thousands from all the marshes and impassable swamps of the north and west ... their flight took place at night ... the clangor of these numerous legions, passing along, high in the air, seemed almost deafening ... and as the vocal call continued nearly throughout the whole night, without intermission, some idea may be formed of the immensity of the numbers now assembled in their annual journey to the regions of the south.

Despite the vivid description, Nuttall probably experienced a flight of sandhill cranes; this account transcends the beliefs of even the most ardent advocates of the "whooping cranes once occurred in flocks" school of thought. This misidentification in itself posed small offense — until Nuttall became an expert.

In 1834, he published his *Manual of the Ornithology of the United States and of Canada: The Water Birds.* This widely-quoted text was so rife with inaccuracies — almost to the point of hilarity — that some ornithologists have since questioned whether Nuttall ever saw a live whooping crane. He describes their habitat as "dark and desolate swamps" and noted that they "build their nest on the ground ... raising its sides to suit their convenience so as to sit upon it with extended legs." One can only imagine a statuesque whooping crane straddling a towering mound of vegetation. Furthermore, Nuttall erroneously suggested that whooping cranes occur in all parts of North America, including the West Indies, where they allegedly spent their winters.

Nonetheless, Nuttall's reputation as an eminent ornithologist spread like wildfire. Within 10 years of the publication of his ornithology manual, friends William Gambel and John James Audubon had commemorated him in the taxonomic names of three new bird species — Nuttall's woodpecker (*Picoides nuttallii*), common poorwill (*Phalaenoptilus nuttallii*) and yellow-billed magpie (*Pica nuttalli*). His name also adorned many plant and mammal species, and his manual was reprinted well into the next century. Fortunately, reason gradually replaced whimsy.

Thomas Nuttall was not the only 19th-century ornithologist to succumb to the mysterious whooping crane. Alexander Wilson arrived at this station vicariously. Wilson was an apprentice weaver in Scotland when he was encouraged to emigrate to the United States in 1794. He had been recently released from a short jail term for writing satirical poetry that revealed unfair treatment of weavers by their employers, which allegedly incited discontent among local Scottish workers. Anxious for a fresh start in the New World, Wilson took up a life of teaching and art. In an 1803 letter to a friend, he declared: "I have had many pursuits since I left Scotland ... music, drawing, etc. etc. I am now about to make a collection of our finest birds." Wilson spent the next two decades roaming the wilderness, compiling observations and painting wildlife.

Wilson saw whooping cranes on the Waccamaw River in South Carolina, on Kentucky ponds, and in the salt marshes near Cape May, New Jersey. He contributed significantly to the bank of whooping crane knowledge — their habitat, behavior and distribution — all of which was published in his 1829 book *American Ornithology*. Wilson, however, made one significant error that would later contribute to considerable confusion. He suggested that the "brown crane" (*Ardea canadensis*) was merely a young whooping crane albeit admitting that he had never seen one. Fortunately for Wilson, the editor of *American Ornithology* corrected this statement after seeing a specimen in Philadelphia's Peale Museum. A footnote on page 86 reads: "It is known to travelers by the name of Sandhill Crane."

John James Audubon was not so lucky. Audubon is frequently considered the premier 19th-century American ornithologist and natural history painter. His interest in birds began in Pennsylvania shortly after he had emigrated from France in 1803 to avoid the Napoleonic Wars. There, he conducted the first known North American banding experiment by tying strings around legs of eastern phoebes (*Sayornis phoebe*) to determine if they used the same nesting sites each year. Audubon eventually settled in Kentucky, where he ran a dry goods store, all the while amassing a large portfolio of drawings and paintings. When his business venture failed, Audubon took his gun and his paintbox and traveled through the United States searching for subjects to paint. His most famous work, *Birds of America*, was originally published between 1827 and 1838 as a double elephant folio, a large picture book containing 435 hand-colored, life-size depictions of North American birds intended for sale to the wealthy of Great Britain. Victorian Europe was embroiled in a love affair with the American wilderness, and Audubon became an overnight success. *Birds of America* contains two paintings of cranes.

Audubon likely saw his first whooping crane near Louisville, Kentucky, on March 20, 1810, while in the company of Alexander Wilson. He started painting Plate 226 — the Hooping Crane — in New Orleans in 1821 but completed the work with the addition of baby alligators and other background elements by 1834. Before painting his subjects, Audubon typically killed them with fine shot, then used rigid wires to place the birds in natural poses. By adding realistic habitat backgrounds, his work appeared to be drawn from life, quite unlike the two-dimensional representations of contemporaries such as Wilson. However, Audubon's second Hooping Crane (Plate 261) was painted from a live bird captured on the Florida coast by US Navy Captain Clack, commander of the war sloop *Erie*. Audubon identified the bird as a juvenile whooping crane "changing from greyish-brown to white" and painted it with white adult cranes in the background. But Audubon was wrong; this bird was a sandhill crane.

Unfortunately, Audubon's errors go well beyond the two paintings. His inability to correctly identify the two cranes has been a great detriment to those attempting to reconstruct historical distributions. Audubon traveled extensively through Kentucky, Mississippi, Louisiana, Florida and Texas when the region was still relatively untouched by colonists, yet we must consider much of his data unreliable.

Audubon deserves mention in the whooping crane's story for another reason. Many wildlife artists of his time shot their subjects before painting

John James Audubon, renowned ornithologist and natural history artist, traveled throughout the United States painting birds from specimens that he collected in the wild.

them, but Audubon's approach somehow better reflects that unconscious, cavalier killing of wildlife that began in North America with the arrival of the European settlers. Audubon himself once wrote: "I call birds few when I shoot less than one hundred per day." One biographer would disclose, "The rarer the bird, the more eagerly he pursued it, never apparently worrying that by killing it he might hasten the extinction of its kind." Furthermore, Audubon's 1840 text describes one encounter with whooping cranes:

> I had so fair an opportunity that I could not resist the temptation …
> I felt confident I must kill more than one … I fired. Only two flew up,
> to my surprise. They came down the pond towards me, and my next shot
> brought them to the ground. On walking to the hole, I found that I had
> disabled seven in all.

One would wonder how many birds it is really necessary to kill in order to paint one's portrait.

We will never be sure how many whooping cranes graced the grasslands and marshlands of North America when the settlers first arrived, but they were seen regularly during the early to mid-19th century from the Arctic coast to

central Mexico and from Utah to New Jersey and South Carolina. In some ways, the question of their abundance in 1870 is purely academic. Once the plains were settled, whooping cranes would only barely survive the next half-century. They may have numbered many millions for what it would matter; humans have demonstrated an astonishing propensity for wholesale destruction of wildlife despite its abundance. Passenger pigeons (*Ectopistes migratorius*) may have been the most common birds in the world, numbering an estimated five billion individuals. Their densely packed migrating flocks were over 1 mile (1.6 km) wide and 300 miles (482 km) long, and took many hours to pass overhead. However, within two decades of unregulated hunting — from 1870 to 1890 — the passenger pigeon was pushed to near extinction. The nail in the species' coffin was driven in April 1896 near Bowling Green, Ohio. Hunters had promoted the event to their compatriots using newly erected telegraph lines; it was one final chance to shoot the last flock of passenger pigeons, and they knew this well. They descended on the Green River nesting grounds en masse and, in a few hours, killed 240,000 adult pigeons, leaving 100,000 nestlings to perish; only about 5,000 birds escaped. But even then, the killing did not end; in 1900, the last wild passenger pigeon was shot by a 14-year-old boy on March 24 in Pike County, Ohio. So, whether whooping cranes once numbered 2,000 or 200,000 seems of little consequence; they would be no match for what would come.

By the middle of the 19th century, economic and political events had set the stage for the westward migration of settlers into the Great Plains of the United States and Canada. Vast tracts of land had been incorporated into the Union with the Louisiana Purchase of 1803 and Texas annexation of 1845. These lands had been mapped, surveyed and sectioned many times over. Railroads crisscrossed the plains and steamboats navigated the great rivers. At first, settlement did not extend beyond the forests east of the Mississippi River; the prairie soil was thought infertile and impossible to cultivate. However, the invention of the steel-bladed plow coupled with the testimonials of more adventurous souls convinced the faint of heart that great prosperity was ripe for the taking on the open prairie. The 1869 *Minnesota Guide: A Handbook of Information for the Travelers, Pleasure Seekers, and Immigrants* hails the virtues of a Midwestern life. It promised bountiful timber, minerals and maple syrup, and land so rich that manure would not be required for 20 to 25 years. Furthermore, Minnesota boasted the "healthiest climate in the world," based on the "testimony of thousands of cured invalids." Prime agricultural land sold for $5 to $10 an acre, with easy payment plans and long credit terms available. Furthermore, any Minnesota man over 21 could obtain 160 acres (65 hectares) from the local land office for a mere 12 bucks.

In Canada, settlement of the prairies began in the 1880s when the country's vast distances were finally spanned by the Canadian Pacific Railway. By the Dominion Land Survey, the Canadian government had divided most of

Audubon's book *Birds of America* contained two plates allegedly depicting "Hooping Cranes." Although Plate 226 (shown here) was, indeed, a whooping crane, Plate 261 is more correctly identified as a sandhill crane.

the west into square-mile (2.5 square km) sections. Under the plan, quarter sections were available to willing settlers who could pay the $10 fee. Provided that he had cultivated 30 acres (12 hectares) and built a house, the homesteader received title to the land in three years.

The same story could be told throughout the Great Plains from the prairie provinces south to Texas and Louisiana, only the crops differed somewhat — wheat predominated in the north and corn was grown farther south. But they both meant the same thing to whooping cranes: native grasslands bent to the plow and wetlands were drained to increase the acreage. The devastation was epidemic. Since colonial times, 740 million acres (300 million hectares) of North American grasslands have been converted to agriculture, and 180 million acres (73 million hectares) of wetlands have been lost. Conservationists have suggested that agricultural expansion through the Great Plains and Canadian Prairies may have caused the demise of 90 percent of all whooping cranes by depriving them of suitable breeding and wintering habitats. Certainly, the westward expansion of humans took its significant toll. However, there is evidence that numbers declined much faster than could be explained by habitat loss, and many prime habitats that were untouched by the plow remained unoccupied by nesting or wintering cranes. It was obvious that whooping cranes were dying more directly at the hands of humans as they stretched their arms to the west.

In 1858 Spencer Fullerton Baird, eminent ornithologist and assistant secretary of the Smithsonian Institution in Washington, DC, declared that the whooping crane had been neglected scientifically. He lamented, "there are none in any of the public museums of the United States, as far as I have been able to ascertain." Baird's wish was granted that same year when Thomas E. Blackney of Chicago presented the museum with their very own whooping crane specimen, catalogued in June 1858 as specimen no. 10384. Although Baird's remarks were not designed to incite catastrophe, what followed was a flurry of skin and egg collecting that greatly taxed the faltering populations. To make matters worse, there was nothing more desirable to collectors than something perceived to be doomed. As the cranes became rarer, the monetary gain to be had by killing them increased dramatically.

During the 19th century, oology — collecting and studying birds' eggs — became a fashionable pastime. Some Victorian ladies and gentlemen chose not to search for their own eggs but, rather, purchased them from dealers who maintained a considerable stock of eggs that they had acquired from egg hunters. Walter F. Webb was the editor and publisher *The Ornithologists' and Oologists' Manual* in 1895. Webb advertised for sale 10,000 various egg sets (all the eggs from a single nest) and about 20,000 other single eggs of various species. Wholesale prices were available on overstocked items. For readers destined for a career in nest robbing, he added that he typically bought from suppliers in quantities of 1,000 to 3,000 eggs.

In the early days of collecting, before any perceived shortage in North American birds, eggs could be purchased at bargain prices. Whooping crane eggs sold in sets of two for 50¢ each. By 1890, however, their price had quadrupled; five years later, they had increased to $3.00 each. Although collectors complained that cranes were becoming exceedingly rare, they continued to take their eggs wherever found, usually shooting agitated parents to facilitate the theft. Ironically, some avid collectors were also enthusiastic ornithologists who heartlessly destroyed nests of birds that they claimed to revere. These Jekyll-and-Hydes produced among the most pathetic accounts to appear in late 19th century ornithological literature; J. W. Preston was a regular contributor. He describes his encounter with a nesting female in Iowa, not too far from where the last whooping crane nest in America was found some years later:

> To my delight she was sitting on her heavily marked drab egg, which lay in a neat cavity in the top of a well-built heap of tough, fine marsh grass … The eggs were the first I'd seen and were a rare prize to me. When I approached the nest, the bird, which had walked some distance away, came running back … trotting awkwardly around, wings and tail spread drooping, with head and shoulders brought to a level with the water … then with pitiable mien, it spread itself upon the water and begged me to leave its treasure, which, in a heartless manner, I did not.

To compound the tragedies, these collected eggs have little scientific merit now. Most of 121 whooping crane eggs known to have been taken between 1880 and 1900 lack date or location data, thus rendering them useless in any historical context. Perhaps the only value that remains to us now is the compassion that may be generated by the description of the mournful parents.

Many wealthy Victorians also enthusiastically acquired whooping crane skins for their private collections. Iowan collector Daniel H. Talbot had 7,000 to 8,000 bird skins in his private holdings that he had purchased for as little as 10¢ each, depending on species and condition. By April 1887, however, Talbot paid $2.50 each for whooping crane skins or $2.00 if purchased in lots of 12 or more. As whooping cranes approached obscurity prices continued to increase, as did the number of birds, destined for collections, that fell by the gun. Within three years, crane skins would fetch a minimum of $8.00 each, at a time when 12¢ would buy a pound of steak or a 5 pound sack of flour. By 1895, whooping crane skins in Walter F. Webb's catalogue retailed for $18.00 each. Their value was second only to those of the now critically endangered California condor (*Gymnogyps californianus*), which could fetch $50.00. In many ways, Webb's catalogue poignantly predicts the tragedies that would befall other birds. Species now extinct or virtually so, such as the heath hen (*Tympanuchus cupido*; $15.00) and ivory-billed woodpecker (*Campephilus principalis*; $15.00), could demand a high return. At least 268 whooping cranes were killed for their skins, skeletons and eggs between 1858 and 1922. This number, however, represents only those birds that were recorded as they passed into private or museum collections. This, in itself, cannot account for the rapid decline in the species during the last three decades of the 19th century. Other things were afoot in those days.

The exquisite beauty that we have come to admire in birds had not escaped the eyes of 19th-century women's fashion. Wearing feathers began in European culture when Marie Antoinette of France donned an elaborate feathered headdress in the 18th-century court of her husband, King Louis XVI. By the 1850s, the burgeoning millinery trade was killing birds on an enormous scale. Although the soft nuptial plumes of egrets were the feather hunters' prime targets, many other white birds — including whooping cranes — were in high demand. Florida and Louisiana were the source of most birds killed for the feather trade. Southern wetlands were opened to marauders as roads were built, and the Texas Gulf Coast became an outpost for trafficking skins and plumes to the cities of Europe and the American northeast. There were no laws to govern these actions; everything was there for the taking. Plume hunters frequently destroyed thousands of birds in their colonial nesting colonies, leaving their eggs to perish and their chicks to die of starvation. Sleeping birds at the roost were mowed down with no more compassion than one would give a bent blade of grass, and their rotting skinned carcasses were left behind to litter the beaches and mudflats.

Hat makers had an insatiable appetite, and millions of North American birds were killed for their feathers between 1870 and 1900, at the height of the feather trade. Ambiguous methods of accountability make it difficult to determine precisely how many feathers left the United States bound for London and Paris markets. It took six large birds to make an ounce of feathers, but in 1911, four companies in London alone reported that their nine-month supply of white feathers was 21,528 ounces. Between 1901 and 1910, ornamental plumage valued at 20 million pounds sterling was imported duty-free into the United Kingdom. Earlier, the March 21, 1888, "Public Sale" inventory of Hale & Son of London listed more birds' skins, wings and feathers than could be found in the combined holdings of all American private and museum collections.

The atrocities that befell the avian world at the turn of the 20th century had one single affirmative outcome — they gave birth to an allied wildlife protection movement in North America. Spearheaded by such eminent names in American ornithology as Joel A. Allen, the American Museum of Natural History's first curator of birds, and Frank M. Chapman, owner and editor of *Bird-Lore* and Allen's successor at that museum, the cause enlisted the aid of scientists, naturalists, authors and concerned laypeople in changing the American public's sensibilities surrounding wearing plumage.

Millions of North American birds met their demise in the name of fashion when Victorian women chose to wear their beautiful feathers on hats.

Key players in this mission were the women of the Audubon Society. Prior to this, the wildlife protection movement was primarily the mandate of men who were only interested in protecting game animals in order to have more quarry. However the Audubon Society — originally established by George B. Grinnell (editor of *Forest and Stream*) in 1886 in the name of the famous naturalist and artist — was committed to conserving all wildlife and habitats through public education and legislation. Terms of membership were quite clear — they required each prospective member to sign and submit their pledge to prevent killing wild birds, particularly nongame species, for sport, eggs, nests or feathers. In 1887, Miss A. C. Knight published her request for an additional pledge in *The Birds Call*, the journal of the Pennsylvania Audubon Society. The members' promise would also read:

> *I pledge myself not to make use of the feathers of any wild birds as ornaments of dress or household furniture, and by every means in my power to discourage the use of feathers for decorative purposes.*

Tides began to turn as wealthy women used their influence with their peers and other prominent society members to abandon the use of feathers in fashion, and to change state and federal legislation. The first law to substantially hinder feather trade profiteers was the Lacey Act of 1900, which prohibited interstate commerce in protected wildlife. Although it did not directly prevent birds from being killed, the Lacey Act stopped feather hunters from transporting their cargo into states, such as New York and Massachusetts, where birds were already protected. Feather merchants and milliners in New York and Boston fought the Act in court, suggesting that thousands of workers would lose their jobs, but the law was not repealed. The Lacey Act's only loophole was that birds could still be decimated for export in states where no legal protection existed. Eleven years later, however, the Audubon Society successfully imposed the Audubon Plumage Bill, which banned the sale of all native birds' plumes, and in 1913, the Tariff Bill prohibited the import of wild bird feathers from other countries.

The American feather trade was shut down, and milliners did not go out of business. Ribbons and lace replaced feathers, and eventually North American women lost the desire to wear birds on their heads. But damage had been done. In 1913, ardent conservationist Dr. William T. Hornaday listed 61 avian families and species that he considered in danger of annihilation for women's fashion. Many had escaped just in the nick of time — perhaps, so did the whooping crane. We have no concrete measure of the plume trade's impact on crane populations except to say that the species was regularly hunted for its feathers; maybe their innate wariness prevented their extermination in this way. Regardless, it was not profit that drove whooping cranes through that appalling bottleneck early in the 20th century, it was sport.

The Victorian era, and the two decades that followed, were also the years of no bag limit, and enthusiastic sport hunters demonstrated the same arrogant disregard for other living beings as did those seeking valuable plumes. The body count was astronomical; in 1909 alone, hunters killed 5,719,214 game birds in Louisiana. Moreover, these horrific numbers would not account for the mass destruction of nongame birds — meadowlarks, woodpeckers, hawks, owls — that were taken in target practice. Birds were often killed when at their most vulnerable, at the nest or when roosting. During a 1910 "torchlight" bird hunt along the Trinity River in northeast Texas, 10,157 sleeping robins were shot in two hours and ten minutes. These birds were killed neither for food nor for any commercial gain; it was slaughter, not sport.

Gun clubs were popular among the wealthy in the late 19th century; bracing autumn weather surely spurred humanity's killing instinct among these otherwise gracious Victorian gentlemen. In Midwestern flyway states such as Nebraska, hunting contests were held semiannually to determine which subset of members shot with the greatest proficiency. On the 22nd of September 1886, the Omaha Gun Club placed an advertisement in the *Omaha Daily World* titled "The Gun Club Proposes to Go Gunning for a Banquet":

> *The Omaha Gun club held a meeting last night and instructed the secretary to issue a challenge to the Omaha Sportsmen's club for a club hunt, the losers to pay the expenses of a banquet. The revised Omaha Gun club rules governing the count of game will be used. Snipe count from 1 to 8, geese from 8 to 12, cranes 15 to 20, herons 10 to 15, swans 30, ducks 4 to 8, eagles 25, and miscellaneous from 3 to 25.*

Game rules were simple — kill as much as you can in one morning outing. Point values attributed to each species would be totaled for each side to determine who would host the banquet.

The autumn contest of 1886 was legendary. Dr. H. A. Worley of the Omaha Sportsmen's Club was captain of the winning team, which made a total score of 1,503 points (327 birds); the losing team, captained by J. J. Hardin, scored 1,468 points (233 birds). General G. S. Smith received special commendation for killing 78 birds himself (322 points), including 13 hawks and an owl. The biggest losers, however, were not the members of the Omaha Gun Club, but the 194 blue-winged teal ducks that met their demise that crisp autumn morning. The lavish banquet was held on October 23 at the Millard Hotel. The defeated toasted the victors by gaslight until midnight at tables heaped with food and drink served in glittering crystal and silver vessels. Good cheer and camaraderie flowed like oaky wine. And decoratively suspended above the head table was the slaughtered body of a whooping crane – the trophy of the hunt — which had conferred an impressive 20 points.

They called them the *bugle cranes*, and hordes of sport hunters sought them out wherever they could be predictably found. The greatest numbers of whooping cranes were shot while migrating through Nebraska, particularly during the 1880s and 1890s. Their traveling routes were well mapped, and the importance of the Platte River valley to staging and roosting cranes had been known to Europeans since they traveled the Oregon Trail in 1841. Fully two-thirds of all whooping crane sightings occurred along this waterway. Hunters longed to kill something that was so big and beautiful, and the allure of the challenge was intoxicating to them. They deliberately declined easier opportunities to kill smaller, grayer, more plentiful sandhill cranes if the

North American birds had little protection from the whims of sport and market hunters until the Migratory Bird Treaty Act was ratified in 1916, making it illegal to harm nongame species.

possibility to bag a whooper was present. Albert M. Brooking, founder of Nebraska's Hastings Museum, looked back from 1943 on the decimation:

> *[They were] killed in central Nebraska each spring, for many hunters in those early days killed about everything that crossed their paths, more to test their skill than for game as the Whooping Crane were never considered good eating. The birds were wary and did not decoy, but their large size made them an enticing target that few could resist when they came into range.*

Although many hunter-authors lamented their acknowledged "wickedness" in the relentless pursuit of whooping cranes, the killing continued. And as the cranes became rarer, the frenzy to kill them was augmented.

Theodore S. Van Dyke, an avid hunter, needed only two shots to knowingly edge the species closer to oblivion. He captured these late Victorian sensibilities in his 1895 book *Game Birds at Home* as he described how he neatly dispatched a mated pair:

At the report of the first barrel, one with folded wings and drooping neck turned its course into a downward plunge, and with the second another relaxed its hold on the warm sunlight and ... descended in a revolving whirl of white, black and carmine.

Thus, migrating cranes by their nature were highly susceptible to devastation by hunters. Dependent young birds were unable to survive without their parents; when one member of a family group was killed or wounded, others remained with the downed bird, making them equally vulnerable. Entire families were frequently destroyed.

Hunting tales of this vintage are rife with accounts of wounded cranes with crippled wings or legs that were typically left to die. To avoid wet feet, bodies that plummeted into water or marshy vegetation were not retrieved. Dead cranes also adorned shop windows, often eight or ten at a time, in cities as far away as Chicago. Their pure white plumage attracted admiring attention of passersby, perchance to lure these potential buyers inside. Unfortunately, no one stopped the carnage to count the dead until it was almost too late. After the 1890s, the record kills by the gun had declined because there were fewer birds left to kill. Whooping cranes, so numerous during migration in 1870, were considered rare in 1890; and by 1901, great flocks of many species were dwindling steadily. Within 20 years of unregulated hunting, even sandhill cranes had been reduced to "thin and straggling bands and white cranes [seen] only in the most isolated instances." One hunter added that he would need a 13-inch mortar to bag his limit of ducks in these days.

These tales speak only of decimation during migration, but whooping cranes were no safer in Louisiana and Texas, where they predictably spent several months each winter. In the distant past, their greatest wintering concentration was likely tallgrass prairies of southwestern Louisiana; these birds would have nested in similar habitat farther north in Iowa, Minnesota, North Dakota and eastern parts of their Canadian range. Louisiana was somewhat more humid than Texas, and allowed greater possibility for lush vegetation and standing freshwater habitats. The almost two million acres of tallgrass prairies that existed in Louisiana before settlers arrived would certainly have supported many thousands of wintering whooping cranes.

Although explorers began destroying the Louisiana prairie wildlife in the 1700s, the greatest devastation occurred some decades later with the arrival of hunters and settlers. One of the last herds of bison was annihilated near the Arkansas-Louisiana border in 1808; there is no reason to believe that whooping cranes would not have suffered the same fate. Ranchers came first to settle southwestern Louisiana. Their hunting parties rode on horseback and encountered the *grue blanche* in great numbers all winter. In 1881, the Louisiana Western Railroad pushed west to the Texas border and the burgeoning towns of Lafayette, Crowley and Jennings received thousands of settlers anxious to cultivate the rich prairie soil and level land. Grasslands that once teemed with waterfowl and wading birds were replaced by farms.

By 1887, rice had become an important commercial crop in Louisiana, and the industry's growth was equally paralleled by the decline in whooping cranes. Within a count of mere years, they were extinguished from Louisiana as they were thoughtlessly killed for sport and meat and as perceived competitors. Fate struck one winter day in 1918 when farmer Alcie Daigle from Cameron County was harvesting his crop. He spotted a flock of whooping cranes that were feeding on rice that was falling from the loose separator door of his threshing machine. In anger, he fetched his gun and shot all 12 birds as they lingered too long around their fallen kin. This dozen were among the last of the Louisiana prairie cranes.

Although reasons may have differed somewhat, these tales of decline could also be told in Texas. This state once had a large wintering population spanning the Gulf Coast south of Louisiana. Cranes were considered to be relatively common in the 1870s and 1880s, but soon fell in great numbers to hunters who often traveled some distance aiming to bag a white crane. Authors claimed that they were a "favorite game fowl" in Texas, and the myth of their abundance spread like wildfire. By 1889, however, they were exceedingly rare. Still, 20 whooping cranes were shot between 1889 and 1904 along the coast from Sabine Pass to the mouth of the Brazos River, only to become museum specimens; these 20 skins represented most of what was left of that population. Moreover, as the handful of remaining whooping cranes held tenuously to their existence, the Texas salt marshes and mudflats to which they had migrated for millennia were slowly being poisoned by oil and sulfur industrial development. Nevertheless, much of Texas's pristine wintering habitat that still lingered in the early 20th century was devoid of whooping cranes. There were no longer enough young birds being born to the north to fill their traditional territories.

Nineteenth-century ornithologists knew very little about the private life of whooping cranes, despite their conspicuous appearance and behavior. It was not until 1874, when Elliot Coues, founder of the American Ornithologists' Union, published *Birds of the Northwest*, that a rational description of the species' breeding range was compiled. Coues considered his own observations, as well as reasonable opinions of others, within the confines of "good biology." Furthermore, he was well aware of the confusion between the two North American crane species that dated back to Wilson and Audubon. Finally, the myths were being relinquished: whooping cranes do not breed in New England and they are not common in Florida, nor do they migrate to South America.

The irony remained in that as quickly as ornithologists could map the cranes' range, the species was disappearing from its traditional distribution. From 1864 to 1922, their breeding range was delineated by a meager 49 nest records in the United States and 7 in Canada. Moreover, at least 31 of these nests had been destroyed by egg collectors who shot the adults to simplify the theft.

Breeding records read like obituaries — no nest reports in North Dakota after 1884; last eggs collected in Minnesota in 1889; only one nest to be found in Iowa by 1894. Illinois, a breeding hub in bygone days, had no records after the 1850s, although cranes likely nested sporadically in that state until the 1880s. Nevertheless, within 26 years of Spencer Fullerton Baird's comment regarding the dearth of whooping cranes in the world's museum collections, the species was within two nests of being lost forever as a breeding bird in the United States. Canada would hold out only slightly longer.

But the end would eventually arrive. Perhaps the last documented nest of the American white crane in the United States was found by ornithologist Rudolph M. Anderson on May 26, 1894, in a marsh south of Crystal Lake in Hancock County, Iowa. Anderson was seeking waterfowl eggs when a local boy told him of a "white crane" nesting nearby. They searched for the nest relentlessly among the rushes and saw grass while the adults circled, attempting to distract the intruders by whooping loudly and dragging their wings. Their nest held two newly laid eggs. Anderson's account fails to mention if he took the eggs; however, he casually admits that he tried to shoot the adults as they paraded around him. Enlightenment would be long in the coming.

All reports of whooping cranes in the United States after the 1890s were migrants en route from their dwindling wintering range to the few northern

Hunters often claimed that they were unable to differentiate between sandhill cranes (above) and whooping cranes (below), and, as a result, whoopers fell regularly by the gun.

breeding sites that still existed in Canada. But even migrants were disappearing as decimation in Louisiana was being reflected farther north. The last record of migrating whooping cranes in Illinois occurred in 1891, and Iowa in 1911. Migrants passing through Minnesota survived until 1917, when two cranes were shot by a farmer in Roseau. When the tally was finally settled decades later, biologists determined that over 90 percent of the world's whooping cranes had disappeared between 1870 and 1900 due to habitat loss through agriculture; egg, skin and plume collecting; and hunting. The remaining birds would dwindle steadily in the decades that followed.

At the turn of the 20th century, however, no one was sure how many whooping cranes were left. Some ornithologists clung to the myth that they were abundant somewhere else; their historical distribution was large and many regions within it were poorly explored. Also, reports of migrants were still occurring; perhaps the birds were becoming more wary of humans as they were increasingly persecuted. Unfortunately, there was little information to refute these arguments, and the resulting confusion delayed intervention. A few ornithologists warned of impending extinction, however. Edward Howe Forbush was the Massachusetts state ornithologist and founder of the local Audubon Society chapter. In his 1912 *History of Game Birds, Wild-fowl and Shore Birds of Massachusetts and Adjacent States* he wrote:

> *The whooping crane is doomed to extinction. It has disappeared from its former habitat in the east and is now found only in uninhabited places. Only its extreme watchfulness has saved thus far the remnant of its once great host.*

Forbush was not far off the mark. By 1912, there were only 80 to 100 whooping cranes remaining in the world. Still they were being destroyed without consideration — 36 birds were known to have been shot between 1912 and 1918 in Saskatchewan, North Dakota, Iowa and Nebraska. The tales differ little from those of the previous century. In October 1912, Henry Clarke and a companion named Quick discovered a migrating family group of five cranes at Wood Lake in Cherry County, Nebraska. They killed four of the five birds, including one juvenile female. The family members were stuffed and placed in private collections in Lincoln. A few days later, two more cranes were shot near Grand Island by hunters who claimed to have mistaken the 5-foot-tall, long-legged birds for geese.

More appeals for sanity appeared in the literature, and William Hornaday spoke the loudest. Hornaday may have been the first true conservationist. He had begun his career as a taxidermist at the United States National Museum, now the Smithsonian Institution, but became chief director of the New York Zoological Park (later known as the Bronx Zoo) in 1896. Hornaday devoted his life to the protection of wildlife, publishing hundreds of books, articles, reports and bulletins advocating immediate cessation of the ongoing slaughter. In 1913, he established the Permanent Wild Life Protection Fund and raised $105,000 to endow it. That year, he penned *Our Vanishing Wildlife*, perhaps the first book to portray an unvarnished account of the tragedies befalling North American birds and mammals. His entry on whooping cranes read:

> *This splendid bird will almost certainly be the next North American species to be totally exterminated … We will part from our stately* Grus americana *with profound sorrow, for on this continent we ne'er shall see its like again.*

Hornaday was actively concerned with the species' survival. He had acquired a wounded whooping crane for the New York Zoological Park and was anxious to find it a mate in order to establish a captive flock. Unfortunately, his appeal in the 1916 *Bulletin of the American Game Protective Association* remained unanswered. Nonetheless, Hornaday continued to be a primary defender of wildlife. He was a critical player in shutting down the feather trade; furthermore, he strove to establish responsible bag limits and restrictions on open seasons, and prohibition of automatic weapons for hunting waterfowl. Most importantly, Hornaday's outspoken nature and compelling arguments substantially influenced government agencies to pass comprehensive legislation that would provide extended protection to North American wildlife.

The law that saved countless wild bird species was just around the corner. In 1916, the Migratory Bird Treaty Act was ratified between the United States and Great Britain to provide much-needed protection for over 800 bird species that migrated between Canada and the United States. This legislation made it unlawful to hunt, capture or kill, sell or purchase, possess, transport, import or export any migratory bird (alive or dead) and its eggs, nest, or products (such as feathers) derived from it. These activities were punishable by seizure of not only the birds but also all personal property used in killing or capturing, as well as fines and imprisonment. The Migratory Bird Treaty Act was comprehensive, and it had teeth. It ended any remnants of the North American feather trade that had eluded other legislation and finally slowed the bleeding that was ravaging wild bird species across the continent. Unfortunately, it did not entirely stop the hemorrhage. The Act became law in 1918, making it illegal to shoot or capture whooping cranes, and certainly fewer met their demise than before, but it was still occurring. Moreover, fear of prosecution caused hunters to bury unreported dead to avoid penalty. By 1918, there were only about 50 or 60 whooping cranes left in existence. At least 25 more fell to the gun in Nebraska, Kansas and Saskatchewan between 1918 and 1922.

Meanwhile, Hal G. Evarts, a regular nature writer for the *Saturday Evening Post*, was doing his best to raise concern. Evarts had spent a life outdoors, writing hundreds of articles, short stories, books and screenplays about the American wilderness, and he had been watching the decline of whooping cranes through an author's eye for decades. He knew that cranes had been shot year after year as they migrated over the salt marshes near Hutchinson, Kansas. He knew a live-bird dealer from Pretoria, South Africa, had offered a local man $3,500 for a pair of crippled birds. The cranes were eventually sold to a New York City dealer for $1,900, destined for Sydney, Australia, but they, too, died in captivity before they could be shipped. Reality rang true for Evarts in 1922 when he acquired a mounted whooping crane skin. The presenter had been tipped off to a dead crane's whereabouts by someone who had watched a hunter kill the bird and leave it abandoned in the mud. The bird was stuffed and given to Evarts, who then donated it to the Yellowstone National Park Museum. Evarts was touched, and his July 14 1923 *Saturday Evening Post* article reads like an obituary. He said, "the whooping crane, perhaps the most majestic bird of all our feathered hosts, has traveled the long trail into oblivion." He was convinced that extinction was imminent.

In 1926, renowned ornithologist Arthur Cleveland Bent, author of the 21-volume series *Life Histories of North American Birds*, echoed Evarts's sentiments — "One of the grandest of North American birds … [was] on the verge of extinction." That same year, John C. Philips reported to other delegates at the Sixth International Ornithological Congress in Copenhagen that there were probably less than a dozen pairs of whooping cranes still alive. No one would paint a bleaker picture.

While others were already writing the cranes' epitaph, Dr. Myron H. Swenk, a professor at the University of Nebraska in Lincoln, was preparing to play his hand in their tale. Swenk was an agricultural entomologist by education, but his keen interest in birds led him to become one of the Nebraska Ornithologists' Union's founding members in 1899 and president of the Wilson Ornithology Club from 1918 to 1919. Swenk was also editor of the journal *Nebraska Bird Review* during the 1930s, which published annotated lists of Nebraska birds and semiannual tallies of migrants passing through the state.

This dreadful era for whooping cranes was also a time of phenomenal growth in popularity of one of North America's favorite pastimes: bird-watching. Fifty years earlier, Wells Woodbridge Cooke, a biologist with the Division of Economic Ornithology in the US Department of Agriculture, had encouraged the public to experience the wonder of birds by observing them, not shooting them. Ordinary people — men, women and children — took to the countryside, providing information on bird distribution and abundance like never before. Swenk knew the inherent problem with amateur reports when he wrote, "no one can be more cognizant of the fact that such records as are here presented must inherently carry a certain amount of uncertainty." However, he was an optimist and, moreover, he was looking for reassurance that the whooping crane was not lost forever.

The reports came pouring in. Cyrus A. Black, whom Swenk considered an "able and wholly reliable field ornithologist," saw two whoopers flying north among a flock of sandhill cranes near Kearney, Nebraska, in the spring of 1920. Twelve days later, he reported seeing 56 whooping cranes at the same locality. In 1923, Henry Williams observed 19 individuals in Walsh County. A 1925 report described a flock of 65 staging on the Platte River. All told, Swenk's records included 73 whooping cranes in 1920, 40 in 1922, 77 in 1925, 115 in 1926, and an earth-shattering 257 birds in 1930. Swenk concluded that, despite losing some cranes to Nebraskan hunters since the Migratory Bird Treaty Act had beeen passed, the law was generally allowing the species to recover. But during the 1920s, Swenk was spending less and less time in the field himself. It never occurred to him that the increase in whooping crane reports might have been due to more eyes watching the skies or, worse, that the encouraging reports were merely misidentifications. Any reports of large flocks of cranes or large white birds — pelicans, gulls, snow geese — were accepted without verification. Everyone wanted to see a whooping crane, and so everyone was seeing them everywhere.

Swenk was a well-respected ornithologist, and his reports dumped a bucket on the flame of frenzy that had surrounded the whooping crane. Even T. Gilbert Pearson, National Association of Audubon Societies president, was convinced that there was nothing to worry about. His 1932 Wilson Ornithological Club address reported that, notwithstanding what John C. Phillips had said at the 1926 International Ornithological Congress, "recent information contributed by Professor M. H. Swenk and others indicates that today [whooping cranes] are somewhat more numerous" than everyone had feared. We can all wish that Myron Swenk had been correct, but the truth was that in 1918 there were only three or four small populations of whooping cranes left in the world, and all of them could be found in Louisiana or Texas some time during their year.

Not long ago, the broad marshes, lagoons and beaches and inland prairie grasslands that stretched from New Orleans south to the Rio Grande delta had supported most of the whooping crane world population throughout the winter. By the 1930s, however, the coastal habitat had been severely degraded by draining wetlands and dredging shallows, primarily for industrial use. Farther inland, most isolated grassland pockets had been lost to cropland or overgrazing. Although usable winter habitat had been significantly reduced, sadly, not enough whooping cranes were being born in the north to replace those that died by the gun en route, and the survivors could not fill what little winter habitat still remained.

Whooping cranes declined steadily through the early 20th century, and by 1941 only 16 birds — 14 adult and 2 young — wintered in the relative security of Aransas National Wildlife Refuge.

Southwestern Louisiana's sea-rim marshes and tallgrass prairie were winter homes for the more easterly whooping cranes. However, only a handful had survived through the 1920s following the population's decimation by farmer Daigle in 1918. In March 1928, ornithologist E. W. Wilson was investigating the habits of snow geese (*Chen caerulescens*) in the coastal marshes of Pecan Island when he saw a pair of whooping cranes flying slowly overhead. He lamented the species' loss to the Cooper Ornithological Club the following year:

> *As my eyes followed these birds moving deliberately away a feeling of sadness arose as I realized this was probably my last sight of some of the very few survivors of one of the finest birds native to our fauna but doomed to early extinction. So far as I could learn among the trappers and hunters living in these marshes, from the delta of the Mississippi to the border of Texas, this pair is all that survives of the many Whooping Cranes that once wintered there.*

Wilson added that this lone pair had never brought young from the north. By 1935, whooping cranes would never be seen there again.

Old-timers told of a second group of whooping cranes in Louisiana above White Lake — perhaps, once numerous — that never migrated north in spring but stayed south to raise their young in fresh wet meadows south of Mermentau. Much of this land was covered by 5 to 8 inches (13 to 20 cm) of standing water nearly all year, and was inaccessible to all but the most diligent gunmen. This population's fate was sealed in 1929, however, when the Army Corps of Engineers extended the Intracoastal Waterway through the marshlands from the Vermilion River to Grand Lake and, in the process, "re-found" the sequestered cranes. The waterway provided public access to these almost pristine wetlands and, by the 1930s, only a dozen or so survived near White Lake.

A handful of cranes also wintered in the relative security of the King Ranch in Texas. This legendary homestead had been acquired, in part, by Rio Grande steamboat captain Richard King in 1853. Despite King's impoverished upbringing, he and his partner, Captain Gideon K. "Legs" Lewis, used their ingenuity and entrepreneurial nature to assemble the largest ranch in North America. At its largest, the King Ranch encompassed 1.2 million acres (5,060 square kilometers) — an area larger than the state of Rhode Island — and crossed six counties on the southern Texas Gulf Coast. Whooping cranes had been observed in the area since the 1840s, but it was not until about 1905 that the birds were more carefully noted. About 20 cranes wintered on the King Ranch in 1912. Over the next two decades, the flock dwindled steadily: 16 individuals in 1917, 9 in 1922, 6 in 1927 and 3 in 1933. Some ornithologists suggested that there was less standing water on the King Ranch year after year and, therefore, conditions became increasingly less favorable for the cranes. However, the number of birds never declined in winter, just fewer came back to the ranch each autumn. There were only one or two crane flyways left, and the predictability of migrating birds made it easy for hunters to find them en route each year, despite their rarity. The last two whoopers wintering on the King Ranch departed for their northern breeding grounds in the spring of 1937 and never returned.

The largest group of wintering whooping cranes at this time could be found among coastal lagoons and beaches of the Blackjack Peninsula, which jutted out into the San Antonio and Aransas bays north of Corpus Christi, Texas. In the 1920s, Blackjack Peninsula was winter home to about 25 cranes, most of which likely bred in Canada, where nesting birds still existed. But even in prime breeding habitat such as Saskatchewan, whooping cranes were slowly being decimated.

One of last recorded whooping crane nests was found on May 28, 1922, presumably near Muddy Lake, Saskatchewan. Chief Game Warden Fred Bradshaw had located the nest by watching the behavior of an adult who circled within 110 yards (100 m) as it attempted to distract him. As Bradshaw photographed the nest, he heard a "strange, piping whistle." A newly hatched whooping crane chick — almost the last of its kind — was desperately calling to its parents for food and comfort. Bradshaw quickly snatched up the chick and expertly wrung its neck. It was laid to rest in the possession of wealthy Torontonian James Henry Fleming, who bequeathed his immense private collection to the Royal Ontario Museum in 1940. You may find this small soul on the fifth shelf from the bottom, third cabinet from the back on the left side, with a tag marked "specimen no. 30393" tied around its tiny leg — a testament to the cruelty that nearly drove the species to extinction. No one would see another whooping crane chick in the wild for over 30 years. And although unconfirmed reports in Saskatchewan would continue into the 1930s, Muddy Lake would soon be dry as dust and as devoid of whooping cranes as the Great Plains that stretched off to the south.

Back in Nebraska, phantom whoopers were migrating by the score. By 1933, Myron Swenk had tallied almost 700 spring and 1,000 autumn migrants crossing Nebraska in the previous 20 years — and the numbers were apparently increasing. In 1934, Albert M. Brooking of the Hastings Museum took over management of migration reports. Brooking was located east of where Swenk had collected records, and thus relied on a different set of observers. Nonetheless, these keen watchers reported over 250 alleged whooping cranes during the spring flight of 1934, often in flocks numbering 60 or more. Brooking later pared the list down to 134 "legitimate" sightings, many of them made by old hunters who had no perceptible difficulty correctly identifying a whooping crane so long as they did not have a gun in their hands. But they still made the record books through conventional means; between 1930 and 1940, hunters killed at least 16 whooping cranes and wounded two others.

As late as 1937, Swenk clung to the notion that spring flights averaged at least 100 whooping cranes — observer Black recorded 159 spring reports that year. Swenk knew that small flocks or pairs and trios of cranes were more commonly seen, but he was convinced that these little groups were merely stragglers of the still undiscovered "great flock." Swenk maintained that whooping cranes were not declining, but he did not do so out of malice or imprudence; he was just wrong. The American Ornithologists' Union would later estimate the 1938 whooping crane population at "something less than 300," but even this meager number could not reflect the paucity of birds wintering in Texas, and certainly there were no known breeding sites left that could account for Nebraska's mythically enormous migratory flocks. Nonetheless, hope still lingered that whooping cranes nested or wintered somewhere else.

Swenk's "great flock" never appeared, and the American Ornithologists' Union report had overestimated reality by a full order of magnitude. By 1938, only 11 whooping cranes remained in the nonmigratory flock near White Lake, Louisiana. That same autumn, only 18 birds — 14 adults and 4 juveniles — returned from their remote northern breeding grounds to winter on the beaches of Blackjack Peninsula. There were no others; the Louisiana migrants that wintered near Mulberry Island had expired in 1935 and the last of the King Ranch population vanished two years later.

A critical turning point in the whooping crane's plight occurred in the late 1930s — just before the curtain came down. The ravages of the Depression had left Washington primed to spend money in order to create jobs and kickstart the economy. In addition, the alarming depletion of wildlife resources over previous decades had become increasingly apparent. For the first time in history, government turned its eyes toward conservation, with the purchase of hundreds of thousands of acres of land specifically for wildlife refuges. Biologists were dispatched in all directions from the Bureau of Biological Survey (now the US Fish and Wildlife Service) in Washington to search for possible locations to establish sanctuaries; Neil Hotchkiss was one of them.

When Hotchkiss visited the Blackjack Peninsula during the winter of 1936, he was impressed with the multitude of birdlife that relied on the mudflats and tidal inlets on this stretch of coastal Texas. During his stay, he saw four whooping cranes. He reported his findings back to Washington, noting this ideal location for a nature reserve. Hotchkiss returned to Blackjack for a second look accompanied by two colleagues: J. Clark Salyer II, chief of the Refuge Branch, and Dr. George B. Saunders, an ornithologist for the Bureau of Biological Survey. Saunders had studied cranes in Africa some years before and well appreciated their beauty and ecological significance; moreover, he was distressed at how few whooping cranes were feeding on the plentiful blue crab and killifishes on the brackish "east-shore flats" between the freshwater ponds and the sea. What was worse, he found them nowhere else in Texas.

He recommended the purchase of Blackjack Peninsula and two protective barrier islands — Matagorda and San José — located just offshore, where he had also seen whooping cranes foraging. Barrier islands, by nature, protect the coastline from erosion during storms, and would provide the peninsula refuge with an added degree of privacy from offshore traffic. Also, as they form, sand and other debris are deposited in shoreline shallows, and the resulting lagoons silt up to become salt marshes, prized habitat for innumerable wildlife species. Hence, these coastal wetlands, ponds, inlets and mudflats would provide the cranes with a sizable number of winter territories. The live oak (*Quercus virginiana*) groves and sweet bay (*Persea bordonia*) brush covering the low, rolling uplands further inland would offer acorns for food and shelter from inclement weather.

Hotchkiss knew the importance of the entire Gulf Coast of Texas, which extended 360 miles (580 km) between Louisiana and the Mexican border, to hundreds of other bird species that might be found there. So he also recommended purchase of the Hallinan Ranch, a large tract of land adjacent to Blackjack Peninsula. This would also provide optimal grassland habitat for the endangered Attwater's prairie chicken (*Tympanuchus cupido attwateri*): two birds

In winter, whooping cranes relied on the abundance of blue crab (*Callinectes sapidus*) that could be found on the mudflats of the Blackjack Peninsula in coastal Texas.

for the price of one, so to speak. Like the whooping crane, the prairie chicken had been ousted from most of its native habitat when tallgrass prairie was plowed under for cultivation. However, the bureau back in Washington balked at buying that much land in one place; they had been mandated to disperse their funds wider geographically. Saunders was asked to make a choice: what was most important? "The peninsula," he replied. The additional purchase of the two barrier islands and the ranch was turned down. This decision would nearly spell the end for both the whooping crane and Attwater's prairie chicken.

On December 31, 1937, President Franklin D. Roosevelt signed executive order number 77841 to create the Aransas Migratory Waterfowl Refuge (now Aransas National Wildlife Refuge). In the end, the bureau purchased 47,215 acres (19,107 hectares) — almost 74 square miles (191 square km) — of Blackjack Peninsula for $463,500. For this bargain price, however, they failed to acquire mineral rights to the property. These were retained by San Antonio millionaire Leroy Denham, who owned the property during the 1920s. Denham had used the peninsula primarily for grazing cattle, and with its isolated location and restricted access, the cranes had wintered in relative peace and security.

The following autumn, on October 3, 1938, James O. Stevenson was appointed the new resident manager of the Aransas refuge. He requested CCC (Civilian Conservation Corps) and WPA (Works Progress Administration) workers to improve the roads and begin construction of a park headquarters building. The CCC and WPA were initiatives in Roosevelt's New Deal program, designed to overcome poverty and unemployment during the Depression by creating, among other things, construction jobs on government projects. Stevenson was anxious to increase public awareness of the new refuge and had determined that these were necessary improvements to the property. But the first true order of business was to census the wintering whooping cranes to assess the degree of crisis.

Stevenson did not have to wait long. Not three weeks after his installation as manager, a flock of 40 sandhill cranes landed near Mustang Slough on the reserve's northern side. Two whooping cranes were among them. By November 6, eight more had arrived, including a rusty-plumaged juvenile that had survived its first migration. Stevenson set out to count the winter visitors. Since he had no access to an aircraft, and roads through the refuge were yet to be built, he trudged through the salt marshes and mudflats on foot. He found 14 whooping cranes, four of which were juveniles. Recognizing the difficulty of the terrain and the birds' natural wariness, he acknowledged perhaps missing a few individuals, particularly if they had settled on a barrier island. Thus, Stevenson estimated the 1938–39 winter whooping crane population at Aransas to be 18 birds.

Weather was mild and clear the first winter. Hunters came out in droves to duck blinds set up in San Antonio Bay and St. Charles Bay, just beyond the refuge boundaries. They killed 7,806 ducks and 136 geese that season; fortunately, no one shot a whooping crane. Stevenson continued to observe the cranes in relative peace, making notes on their territorial and pair-bonding behavior. Between mid-April and mid-May, they departed for their breeding grounds, somewhere still yet to be discovered in the distant north. Everyone at Aransas crossed their fingers, anticipating how many would come home next autumn.

The isolated peace of the previous autumn and winter departed with the cranes in spring 1939, when scores of people descended on the refuge. CCC workers were busy building roads and buildings and clearing firebreaks. They constructed a spillway for Burgentine Lake to regulate the reservoir's water levels for waterfowl use. Furthermore, the Bureau of Biological Survey had provided the San Antonio Loan and Trust Company, representing Leroy Denham, with a permit to graze 4,000 head of cattle on the reserve at a cost of 30¢ per head per month. San Antonio Loan and Trust had also leased the drilling rights to Continental Oil, whose earlier oil explorations on the refuge's inland edge had shown potential. But amid the cacophony, Stevenson remained focused. In addition to the whooping crane census, his first annual report would record 186 other bird species at Aransas, including the rare Hudsonian godwit (*Limosa haemastica*) and Attwater's prairie chicken. The cranes returned in October. Five pairs of adults were accompanied by young; remarkably, two pairs had twins. Five other adults were otherwise unaffiliated. These birds may have lost their young to predators or hunters, or they may have been too young or too old to breed. Twenty-two whooping cranes — that was all.

When the Aransas refuge was first established, the protected area included a strip of water surrounding the peninsula; however, the hunting ban on this area had never been enforced. Stevenson closed these waters during the 1939–40 season, much to the chagrin of St. Charles Bay Club members, who had placed 33 of their 51 duck blinds along the refuge's edge. Although hunters were compelled to move their blinds back beyond the boundary, this did not prevent the gun club manager from trespassing to drive ducks out of the protected area and past the sights of hunters' guns. Chaos and animosity reigned until the club members realized that the refuge actually increased their chances of killing unsuspecting waterfowl that were unable to quite determine where the safe zone ended.

On January 27, Continental Oil struck black gold at Aransas, on Little Devil Bayou in St. Charles Bay. When the well subsequently caught fire, Stevenson was forced to police not only drill workers and firefighters, but also gawkers from the mainland who came to see the blaze. Unperturbed by the situation and fueled by the success of the first, Continental Oil began drilling a second well.

As if this was not enough to test James Stevenson's strength, the greatest menace of them all arrived in the spring of 1940 — the Army Corps of Engineers (ACE) descended on Aransas National Waterfowl Refuge. The ACE was working its way down the Texas coastline dredging a channel to extend the Gulf Intracoastal Waterway south to Corpus Christi. Here the waterway would be used by barges and other commercial vessels hauling petroleum and chemical products and manufactured goods. Despite protests, the ACE's course did not veer when they got to Aransas. One concerned individual commented, "To say that the Engineers are unmoved by our wishes is putting it mildly."

Rather than weave the waterway through the chain of bays and inlets just offshore, they chose a cheaper route that sliced through the salt marshes so critical for the cranes' survival. As they pushed through the reserve, they drained and destroyed acres of winter habitat where crane families had foraged on blue crab, danced and called to each in unison. Moreover, the dredged material was dumped on areas adjacent to the water, thus destroying even more valuable land. Even worse than the extensive loss of habitat was something else that the ACE brought to shatter the tranquility of Aransas: people. Now a navigable waterway pierced an otherwise inaccessible part of the refuge, and it would be widely used by recreational boaters who would bring guns, commotion and garbage. Robert Allen wrote:

The ditch, nine feet deep and one hundred wide, over a three-hundred-foot right-of-way bisected the very edge of salt flats where whooping cranes had found safety for perhaps two million years. This once isolated tip is no longer secure, or isolated.

When 22 whooping cranes flew north from Aransas in April and May of 1940, they may have assumed that some shred of their previous existence would be waiting for them when they returned in the autumn, but this was not the case by any stretch. During the summer of 1940, Humble Oil began drilling in the bay just beyond the reserve's water boundary, a channel to the oyster reefs was being dredged, and the Air Army Corps had taken up residence on adjacent Matagorda Island.

In its first decades, Aransas National Wildlife Refuge and the animals that it harbored continued to be threatened by poachers, munitions testing, recreational and industrial development and barge traffic on the Gulf Intracoastal Waterway.

Although it hardly seemed possible, life was much worse for resident whooping cranes near White Lake, Louisiana, for they had no sanctuary. While the ACE was tearing up habitat farther south, an early-August hurricane descended on coastal marshes where the meager flock of 13 cranes clung to existence. Innocently named Hurricane Number Two, the storm slammed into Vermillion Parish with maximum sustained winds of about 80 miles per hour (130 km per hour) dumping almost 30 inches (76 cm) of rain on the parish. Extensive flooding drove the whooping cranes inland, and standing water remained 3 to 4 feet (100 to 120 cm) deep over much of their range from August to late October. Only six cranes returned to the coastal marshes after the storm; six others had been killed by hunters and the seventh was wounded by gunshot. Her wing was crippled, but she survived. Now the remaining White Lake cranes were beyond the capability of recovery. If they were still breeding, then no young survived to repopulate the flock. Year by year their numbers dwindled until 1947, when only a lone bird lived near White Lake.

By 1940, only one self-sustaining migratory flock of whooping cranes existed, and it wintered at Aransas. Nonetheless, the American Ornithologists' Union still maintained the myth that this was only part of the greater flock. Workers in the field — James Stevenson at Aransas and John Lynch at White Lake — knew better. The global population hovered around 30 individuals.

Stevenson watched the skies patiently that autumn. The first crane landed at Aransas on October 22, and by December 17 the remainder had arrived — 26 individuals, including four pairs tending five young. He was encouraged that 16 young produced in three northern summers had survived; however, he was worried about ever increasing numbers of "unemployed" birds, those wearing adult plumage but not accompanied by young. Whooping cranes rarely breed before age five, although they appear fully adult by age two. Many of these unaffiliated birds could merely be too young to breed; they would need to survive three or four more years before they could hope to contribute to the faltering species.

The following spring only 23 whooping cranes departed Aransas. One little family failed to migrate and remained in the east-shore flats near Mullet Bay through the summer. Although Stevenson was concerned, this anomaly provided him the opportunity to study the birds' courtship behavior. He watched as adults bowed their heads to one another and performed their high-leaping dances. The male frequently flapped his wings, pointed his head skyward and uttered his trumpeting calls. When the male drove his yearling off the territory, Stevenson was encouraged; perhaps this was a sign that the pair would nest at Aransas. But they did not, and, by August, the trio was together again feeding placidly on the mudflats.

While the cranes danced, a barge moored near a wildcat well just east of the refuge suddenly discharged a quantity of oil into the water. Stevenson was there; he was showing a visitor those eastern marshes when it happened.

His visitor was Robert Allen, who had come to Aransas from the National Audubon Society to study roseate spoonbills (*Platalea ajaja*) that inhabited the outer islands. Time stopped as Stevenson and Allen watched the oil slick travel inland and toward the waterway that would ultimately carry it into the precious mudflats and ponds. Suddenly, the wind direction changed and the oil slick was pushed offshore. Stevenson and Allen drove to Houston to pay a visit to the oil company president. The man was unperturbed. "These things happen," he replied. Another near miss for the whooping cranes.

Southbound cranes returned to Aransas in October of 1941 and, although Stevenson searched intently, he found only 16 birds — 14 adults and 2 young. The species had reached a historic low. To make matters worse, this meant that there were only two or three breeding pairs of whooping cranes left in the world. Stevenson was discouraged; of 23 migrants that had left Aransas the previous spring, only 14 had returned. In his regular report to Washington, Stevenson would write, "We can only wonder about the rest."

Years later, Robert Allen would give us the answer in his 1952 *National Audubon Society Research Report No. 3*. Whooping cranes lived in relative protection in winter at Aransas, and their yet undiscovered remote northern breeding site sheltered birds on the nest. The problem was that great stretch of miles that lay between summer and winter that offered no sanctuary. Despite the Migratory Bird Treaty Act, which made it illegal to shoot whooping cranes, they were still dying by the gun. Between 1938 and 1948, 39 adult cranes were lost; most were killed by hunters during migration. The death count may very well be much higher because we have no measure of how many dead adults went unreported, nor how many rusty-feathered juveniles were brought down on their first voyage south. Perhaps only a tally of contraband taxidermied skins displayed in the recreation rooms of North America would provide this answer.

In 1941, another singleton was pasted into the annals of whooping crane history. On November 25, a female crane with a crippled wing was brought to the Audubon Park Zoo (not affiliated with the Audubon Society) in New Orleans. This was the bird that had been wounded by hunters during the onslaught of Hurricane Two the previous August. She had been captured by a farmer in Evangeline Parish and given to a Mr. La Haye of Eunice, Louisiana, to nurse back to health. He called his pet Josephine. La Haye thought nothing of the bird until a federal game agent told him that she was a rare whooping crane and, by law, he could not keep her.

George Douglass was the beneficiary. Douglass was a stubborn, arrogant, power-hungry man with an inflated view of his own importance. He had been educated in accounting and law, and was a popular attendee at civic meetings and businessmen's luncheons. Moreover, he had recently been appointed the director of the Audubon Park Zoo. When Douglass received the appointment, the mayor dismissed concerns regarding his lack of knowledge of animals or zoo operations — running the facilities would be largely administrative. No doubt, Douglass gladly received the crippled bird. Among the zoo office files is a tattered card that commemorates Josephine's arrival. It reads: "Josephine … whooping crane … 4 feet tall … wing spread approximately seven feet … adult … pure white … found in rice field in Eunice, LA, 1941 … donated by L. O. La Haye of Eunice, La." Handwritten across the bottom of the card is the word *Priceless*.

Josephine's value to the world should not have escaped the awareness of Douglass, but maybe it did. He placed the crane on public display in a small cage under deplorable conditions. She remained there for the next 10 years, just another white bird on view for an insufficiently interested New Orleans audience. Josephine's card had yet to be played, however. This bird would eventually allow Douglass and the Audubon Park Zoo to control the fate of the entire whooping crane captive breeding program for over two decades. Recovery was still well out of reach.

A patient whooping crane parent shows her young chick how to search for food among the vegetation.

Chapter 3

Recovery of the Whooping Crane:
Population

*I*n autumn 1941, there were fewer than two dozen whooping cranes in existence, and most of the world was oblivious as the birds danced heedlessly closer to the very edge of annihilation. However, after 80 years of blissful ignorance, record low numbers somehow struck a chord of reality among ornithologists that resounded as clear as a bugling call, and finally shook them into wakefulness. The American Ornithologists' Union Bird Protection Report boldly announced that the whooping crane's future was more precarious than ever — a far cry from the vague "less than 300" pronounced three years before. They issued a call for action that implored the Fish and Wildlife Service and the National Audubon Society to take immediate steps to determine the species' exact status and "institute practical measures to forestall its extinction." They added that Aransas National Waterfowl Refuge provided less than secure conditions and that factors contributing to the species' demise on the wintering grounds needed to be redressed.

Unfortunately, a call for action is not a simple task. Very little was known about whooping cranes except for a smattering of winter behavior observed primarily among the Texas flock. The distant breeding grounds remained undiscovered, and the long migration route that united winter and summer was unprotected from perils that continued to gnaw at the species' grip. But the light would not shine on cranes for long. On December 7, 1941, the Imperial Japanese Navy launched a surprise attack on Pearl Harbor and America went to war.

The war changed Aransas considerably. Construction crews departed and the CCC camp closed for good as young men donned uniforms to fight for the cause. But life was by no means quiet, particularly with the Air Army Corps (now called the Army Air Force) perched ominously on Matagorda Island. James Stevenson voiced his concern that low-flying aircraft were disturbing the cranes when he witnessed more than one young pilot buzzing white birds near the beach. He entreated their commanding officers to avoid prime feeding areas. Some complied, some did not. In the spring of 1943, 19 cranes left the relative safety of Aransas to journey north; in July of that year, the Air Force began to use Matagorda Island for target practice.

When the American Ornithologists' Union issued its call for action in 1941 even they did not realize how close the whooping crane was to extinction.

The bombing strip site had been chosen because it was away from dense human populations; no one ever thought of cranes. It regularly hosted high- and low-altitude exercises with live ammunition. Objections raised by the Fish and Wildlife Service were generally ignored, so refuge workers hunkered down and settled in for the long run. Apparently no cranes were killed in the commotion that year — at least no bodies were found — however, more than a few cattle disappeared from rancher Toddie Lee Wynne's half of Matagorda Island.

In 1943, Earl Craven replaced James Stevenson as refuge manager in a periodic transfer that typified the service. Before he left Aransas, however, Stevenson outlined several factors threatening the apparent sanctuary offered to cranes in Texas. Consequences of oil exploration and drilling on the refuge were well understood, as were the activities of the Air Force. Other forms of human intrusion were less predictable. The Intracoastal Waterway was, indeed, providing a convenient conduit for illegal shooting, and canal traffic was increasing steadily as development crept along the Texas coast. Also the expansion of live oak brush vegetation over preferred prairie habitat in core refuge areas was forcing cranes to feed closer to the waterway. Moreover, the cranes appeared to be unafraid of small boats. Perhaps they had become more complacent in the apparent protection afforded by the refuge. There were scant reports of cranes used for target practice within the Aransas boundaries, but refuge workers knew it occurred regularly. Most years, fewer birds left Aransas in spring than had arrived the autumn before.

Yet the war-torn whooping cranes faithfully returned annually to Aransas with their meager broods. Craven knew that they could hold their own with protection from illegal hunting during migration and on the wintering grounds. Adult cranes are long-lived and, typically, have high survivability during their reproductive years; nonetheless, 11 adult cranes, and countless juveniles, were lost in the two years between 1943 and 1945. Hunting was difficult to monitor since cranes were now protected under federal and state law; few sportsmen were willing to report even an unintentional kill. By 1945, the global whooping crane count had dropped to 21: 18 birds wintered at Aransas, and only 3 individuals remained at White Lake.

Craven had his difficulties tallying cranes that winter. The birds continually moved from place to place; many were scattered broadly up and down the coast beyond refuge boundaries. Perhaps they were unduly disturbed by activities on Matagorda or Continental Oil's incessant drilling operations. It was equally likely that drought conditions at Aransas had so decreased water levels in salt marshes that the food supply was insufficient to support this small flock. Even more distressing was that an increasing number of wintering cranes sporting adult plumage were not accompanied by young. At this rate, recovery seemed decades away if it was even possible. In the spring of 1945, the refuge fatefully emptied of whooping cranes. That summer another event would crystallize the call for action.

On August 27, 1945, a violent hurricane slammed into the Texas Gulf Coast at Port Aransas and skirted north. Some reports claimed sustained winds of 130 mph (210 kph) and a 15-foot (4.5 m) surge as it passed Matagorda Island. Roads, bridges and oil facilities were heavily damaged, and there was considerable loss of crops and livestock. The whooping cranes, of course, were safe in their northern home, but the storm illuminated what many people feared all along. The species was clinging desperately to the thinnest thread; any isolated stochastic event — storm, drought, disease or disaster — could push this small population into oblivion.

At this time, Dr. Ira N. Gabrielson was director of the Fish and Wildlife Service; his chief of research was Dr. Clarence Cottam. Gabrielson was an ardent wildlife conservationist who believed that America's protected areas should be "refuges in fact as well as in name." However, he served during an era of ever diminishing resources, first the Depression and now the war. The service lacked requisite funding to support special projects such as saving one species, so Gabrielson turned to the National Audubon Society for help.

The society had been established a half-century earlier to contest the devastation of North American birds from hunting for meat, plumes and trophies. They took the name Audubon in recognition of the famous painter-naturalist, but they were, otherwise, not affiliated with John James Audubon. Throughout the early decades of the 20th century, the National Audubon Society and its state chapters had been instrumental in passing significant conservation legislation imposing kill limits that virtually shut down the feather trade and other profit-based use of wild birds. By the 1930s, however, the society realized that birds were declining from other factors, such as habitat loss, and that preserving a species was more difficult than just instructing people not to shoot it. Conservation also required knowing something about birds' life history. Pure biological research was relatively rare at this time; most wildlife species were studied to better manage them, primarily to increase populations of "good" huntable species or to reduce populations of "bad" species, such as agricultural pests. Only the Audubon Society was interested in spending money on birds that had intrinsic, not commercial, value. They had recently pursued exhaustive study of three other endangered species — the California condor, roseate spoonbill and ivory-billed woodpecker — and were more than willing to take the whooping crane aboard.

The president of the National Audubon Society in 1945 was a dedicated and serious-minded Bostonian named John Hopkinson Baker. He had abandoned his previous vocations as World War I fighter pilot and investment banker to join the society in 1934. Baker was committed to an ecological conservation approach, and held strong to the belief that "every plant and animal has its role to play in the community of living things ... all are beneficial." With Baker at the helm, the society became a premier environmental conservation organization, attracting to its staff of scientists and educators such venerable names as Carl W. Buchheister, Alexander Sprunt and Roger Tory Peterson.

Baker soon realized that virtually nothing was known about whooping cranes, particularly during the breeding season. No one had seen a nesting whooping crane for decades and their northern breeding grounds had eluded any efforts to find them. Baker decided that a thorough study of whooping biology was in order. He conferred with Gabrielson and Cottam and, later that year, the Audubon Society and the Fish and Wildlife Service allied to form the Cooperative Whooping Crane Project. The primary purpose of this venture was simply to "determine what steps may reasonably be taken toward further protection and restoration of the species." No other terms of the partnership were formalized, an oversight that the society would regret in future.

The Whooping Crane Project sponsored field investigations, population surveys and general life history research, but leading the agenda was the mission to find the breeding grounds. All involved knew the importance of knowing how, when and where cranes breed — critical information for captive propagation projects — but it was also imperative that potential threats from hunting, human encroachment or natural causes on the nesting grounds be assessed. It was obvious that many cranes left Aransas in spring never to be seen again. Precisely how they met their demise would have to be determined; otherwise any efforts to keep the species afloat in winter would be undermined by what happened elsewhere.

The postwar climate regarding wildlife among Canadians and Americans was ripe for a cooperative venture in the spirit of the Migratory Bird Treaty Act. Obviously, the birds spent part of their year in each country; the last known whooping crane nests had been found in southern Saskatchewan, and as humans pushed them to the edge of existence they had retreated north to the corners of their once vast distribution. Consequently, the hunt would occur in remote northern latitudes where summers hosted roasting heat and mosquitoes and blackflies were the size of Cessnas. Roads would be few, groundwork would be slow and rivers made impassible by beaver dams and lodges. Searchers would have to be both dedicated and durable.

The choice candidate for the position was Robert Porter Allen, the Audubon Society's research director. Allen was familiar with Aransas and whooping cranes — having seen them there a few years earlier — and had recently received accolades for almost single-handedly saving the roseate spoonbill from extinction. Unfortunately, he was still in the army. Consequently, the task of finding the northern breeding grounds fell first to Fred G. Bard, then curator of the Royal Saskatchewan Museum in Regina. Bard had been hired as a student preparator in 1925 at age 17. By the 1940s, he was considered among the best-qualified ornithologists in western Canada, and his long interest and devotion to whooping cranes earned him much prestige in the Audubon Society's eyes. Bard offered his services in the search until Allen's return.

Bard chose to first search his own turf thoroughly. Saskatchewan still offered vast stretches of suitable open and rolling pothole-nesting habitat, and

Robert Porter Allen, research director of the Audubon Society, had recently saved the roseate spoonbill (*Platalea ajaja*) from extinction when he set his sights on the faltering whooping crane.

whooping cranes were regularly seen passing through the prairie province. Bard amassed historical breeding and migration records to establish where the birds could likely be found. He also enlisted magazine and newspaper editors and CBC (Canadian Broadcasting Corporation) radio announcers, who pledged to spread the word. Moreover, Bard made arrangements to fly with the USAF Search and Rescue Group to find the birds himself. When flight restrictions and conflicting air force agendas caused this relationship to fizzle out, the Fish and Wildlife Service volunteered pilot Robert H. Smith.

Bob Smith was a flyway biologist and waterfowl specialist stationed at Winona, in southern Minnesota. He was a graduate of Dartmouth College, where he had studied zoology, botany and geology. Smith's position with the service included aerial assessments of waterfowl populations during autumn migration and on wintering and breeding grounds. Together, Bard and Smith covered thousands of miles by plane and car in two spring and summer months of 1945 as they searched for nesting cranes in central Saskatchewan and eastern Alberta. They saw a plethora of ducks, geese and other waterfowl; unfortunately, they never saw a whooping crane.

Bard and Smith returned to their regular duties in the autumn of 1945 and the project was adopted by eminent ornithologist Dr. Olin Sewall Pettingill, Jr. At the time, Pettingill was an associate professor of zoology at Carleton College in Northfield, Minnesota. He was a man with a rather serious demeanor and a dry sense of humor who insisted on wearing black tie on exam days.

Pettingill had taken a year's sabbatical leave from his teaching duties to study whooping cranes. He met with Bard to discuss the previous summer's fruitless activities, and they pored over records and reports in preparation for the following year's search. Pettingill also discovered that there were two whooping cranes in captivity. Josephine, the White Lake survivor of the 1940 Louisiana hurricane, was still at the Audubon Park Zoo in New Orleans. Another crane, Pete (or Petunia), was in Nebraska. In 1936, Pete had been shot by hunters; he had a maimed wing and was blind in one eye. Since his injuries, he had been kept in a pen near the Platte River by Gothenburg Gun Club members, who doted on their "heron" that was easily tall enough to look over the fence, although Pete's occasional raids on the nests of captive geese that shared his enclosure had earned him the name "Old Devil." Perhaps it would be fruitful to get Pete and Josephine together.

Pettingill, however, was anxious to commence his hunt for the nesting grounds in June. So his wife, Eleanor, joined him in Regina and they traveled north to Waterways, in northern Alberta, to rendezvous with the search plane. An appropriately equipped plane failed to arrive until Terris Moore, a pilot and president of the New England Museum of Natural History, flew up from Boston, offering both his private Taylor-craft plane and his services. Pettingill and Moore flew over 21 hours in several forays from Fort McMurray, Alberta, and Fort Chipewyan, on Lake Athabaska, until Moore was obligated to return home. Toward the end of July, pilot Bob Smith arrived to dedicate another 20 flight hours. All told, Pettingill and his team searched several thousand miles of potential crane habitat in northern Alberta and Saskatchewan during the summer of 1946, but they never found the lost whooping crane breeding grounds. Pettingill reluctantly left the project in September to return to his students.

Bob Allen picked up the gauntlet on a cold, gray October day in 1946. His appointment read simply: "to study the life history of whooping cranes wintering at Aransas, trace their migration route, and hunt for the breeding grounds in Canada." The whooping crane world population at this time was 27 birds — 25 at Aransas and 2 in Louisiana — and only three young had arrived with their parents that autumn. Yet the cranes called and Allen answered. He locked up his house in the Florida Keys, packed up kit and caboodle, and moved his family to the Texas Gulf Coast.

Allen was the consummate field biologist, a conservation activist and no stranger to adventure. He was born the son of a schoolteacher and a lawyer in rural mountain country near South Williamsport, Pennsylvania. His liberal-minded parents taught their son to look beyond the boundaries of conformity and instilled him with the importance of fervent dedication to cause. Even at a young age, Allen viewed the natural world with a sense of mystical reality. He spent his boyhood playing truant so that he could run with the deer or fly south with a flock of migrating sandpipers. The die was cast, however, when his high school biology teacher encouraged Allen to join the Junior Audubon Club. He bought his first pair of binoculars on his 16th birthday and became an ornithologist. Allen later attempted college but was disillusioned by the restraints inherent in 1920s academia. In his own words, he remained "an undisciplined nonconformist, incapable of learning many of the graces and determined to find a way of life wherein the kind of shoes you wore and the sort of knot in your tie were of no importance whatever." After leaving college, Allen boarded a freighter bound for Singapore and set off to seek his fortune. Three years and one shipwreck later, he returned to New York with 48 cents in his pocket. But chance would change Allen's life again when he met Evelyn Sedgewick, a Julliard School graduate destined for a career as a concert pianist. Suddenly, Allen needed a job. He contacted Frank Chapman at the American Museum of Natural History, who sent him to see T. Gilbert Pearson, then president of the National Audubon Society. Allen's career with the society began with a trial job sorting books in the basement; four years later, he grew a moustache to look old enough to sit on the board of directors. He had found a life where the nature of his clothing was not important.

When Allen arrived in Texas with his wife, Evelyn, and their two children, they were already outcasts. Not only were they not Texans, they had also come to study cranes, a dubious occupation — so thought the local residents of Austwell, the town just 8 miles from the Aransas refuge gates. Years before, a neighboring landowner had expressed his opinion regarding avian conservation:

> *I hear the government is buying the Blackjack for a pile of money just to protect a couple of them squawking cranes. They tell me they ain't bad eating, but there is no open season on them.*

But Mrs. Allen would not be flummoxed. She wove her wily Julliard ways, played the organ in church on Sunday, and by degrees convinced Austwell's doubting populace that the Allens were not raving lunatic bird lovers with outrageous liberal tendencies. Her husband would recollect perhaps their greatest victory in Texas, just two years after they had arrived. The owner of Cap Daniel's beer parlor — the true cultural center of Austwell — had asked Allen for a photograph of a whooping crane. In order to make room for it among the mementos that covered the saloon walls, Cap unhesitantly ripped down a large lithograph that graced the most prominent wall in his establishment. It had been a picture of the venerable Judge Roy Bean.

Allen began studying cranes in earnest in November 1946. Although he was ever awestruck at their magnificent silhouettes bright white against the sand, he was dismayed by their remoteness. He wrote:

> *I remember that those first two birds seemed very far away — not only in the physical sense. Their arrogant bearing, the trim of their sails, as it were, would intimidate the most brash investigator. I reached our cabin that first night feeling very humble and not too happy.*

Allen soon discovered that it was difficult getting close to the cranes at Aransas. They were not accustomed to people on foot and moved off when they approached. Allen improvised. He enlisted the aid of Bud Keefer, the Aransas refuge manager, to mind the tractor controls as Allen perched atop a shaky old farm wagon, one hand on his binoculars and the other hanging on for dear life. So it was that they regularly surveyed 12 miles (19 km) of oceanfront habitat. Nonetheless, Allen was unsatisfied by his inability to get close enough to make sketches and detailed observations. He had noticed that otherwise wary birds paid little heed to cattle roaming the refuge. Allen was resourceful; he carefully constructed a giant canvas and wood blind in the shape of a bull, and painted it mahogany red to resemble Santa Gertrudis cattle. The beast blind, christened *Bovus absurdus*, worked brilliantly until it attracted the attention of a real bull, who eyed the interloper with some antagonism. Newspaper headlines flashed through Allen's mind — "Bird-watcher Gored by Bull." When the bull hesitated in his advances, perhaps out of confusion, Allen seized the opportunity and quickly retreated back to camp.

However, his difficult work eventually paid off. Allen had learned the nature of every crane and crane family living at Aransas and had given them names based on their location — Slough Pair, Middle Pond Family, North Group and Dike Pair. He had mapped their territories and gained an understanding of the intricate behavior that maintained these boundaries. He had tallied their food supply — particularly the critically important blue crab (*Callinectes sapidus*) — and had determined how these prey populations were regulated ecologically. Allen already knew well the value of wildlife, but these first months with whooping cranes impressed upon him the inestimable worth of every individual. Each was crucially important to the survival of the species — as an individual, not just a numbered bird in a withering population.

Allen began to compile what he would later call "kill records," detailed accounts that tallied both reliable sightings and death records to produce an accurate estimate of what was and what is, and where lost cranes were disappearing. He determined that the greatest annual accountable losses were still occurring during migration from illegal hunting, particularly along the Platte River and in Saskatchewan grainfields. Allen was a great advocate of public education. He knew that if he could make whooping cranes the most famous birds in North America by telling their disquieting story to everyday people — not just the bird-minded community — then he could dissuade even the most stalwart skeptics. Allen began a media blitz along the migratory route from Texas to Saskatchewan that would last for years. He circulated clear descriptions with photographs to prevent whooping cranes from being confused with ducks and geese and, most importantly, sandhill cranes. He recruited feature articles for hunting magazines and farm journals, and urged school children and youth clubs to participate in the recovery. Allen had additional help from Fred Bard in this regard when information material was distributed to 5,500 schools in Saskatchewan, eastern Alberta and western Manitoba and 500 Natural History Society members. In addition, Bard continued his radio and newspaper campaign into the early 1950s. More people knew about whooping cranes than ever before, but whether or not these efforts would pay off had yet to be seen.

In the winter of 1946, Bob Allen added Pete and Josephine, the two captive whooping cranes, to his agenda. First, he went to New Orleans to take a look at Josephine. He admitted that she appeared in good health at this time, but was disturbed at the sight of a captive whooping crane who caught with expert proficiency peanuts, popcorn and bits of hot dogs thrown by visitors. Her small wire cage pitifully displayed a meager white board on which were painted four words in black: *whooping crane — North America*. Allen knew that the captive birds were not getting any younger; they were, perhaps, 15 years old or more. Arrangements would have to be made to get these birds together, and soon. Allen's options were limited, however, by the birds' disabilities. They would have to remain in captivity, but the choices were few. Pete and Josephine could be housed in a large, natural-environment enclosure at Aransas or they could be placed in a zoo with expert care — Chicago's Lincoln Park Zoo, the St. Louis Zoo and the San Antonio Zoo all had good facilities and expertise in breeding rare birds. Although Josephine was already accustomed to the New Orleans zoo, Allen never considered this as an option. He was justifiably unimpressed with what he had seen.

Allen's search for the elusive breeding grounds began in spring 1947. Once again, he and his wife packed up their children and camping gear in the

Whooping cranes relentlessly led their young between Aransas National Wildlife Refuge and their still undiscovered northern breeding grounds despite the precarious nature of their future.

family caravan: station wagon, car and baggage trailer. They left Aransas ahead of the cranes in April and headed north hoping to intercept them along the way. It would be a one-in-a-million rendezvous. They chose North Platte, Nebraska, as their first way station, where Allen's kill records had shown cranes traditionally passing through. A request for information was printed in the *North Platte Telegraph Bulletin* and broadcast over the radio by local station KODY. Wires were sent from Aransas every time a crane pair or family was not seen on their regular turf.

Tips from the concerned public came flying in — over 144 in the first month — and Allen diligently checked out every one. Reminiscent of the Swenk years, hopeful people saw ring-billed gulls, pelicans, sandhill cranes and snow geese in the guise of whooping cranes. Allen tracked false leads, logging almost 4,000 miles (6,400 km) in his car and by air while the cranes mysteriously moved northward. Then, on April 19, he received a telephone call from a state conservation officer; there were cranes on the Mather farm. Allen jumped into his car and dashed out to the farm, only to find pelicans in a pasture. But Farmer Mather insisted; he knew whooping cranes when he saw them! He pointed to the South Platte River. "They went that way," he said. Allen immediately chartered an airplane and flew over the river. Farmer Mather was right; five cranes were resting on a sandbar, and among them was the "North Group" family. They had taken six days to fly from Aransas to the South

Platte River. Encouraged, the Allens left Nebraska and followed leads to the Canadian border, where the trail went cold. The eerie silence was foreboding. The cranes were traveling the most dangerous part of their journey, where Saskatchewan farmers were particularly hostile to birds in their grainfields. The Allen caravan continued north to pick up the search where Pettingill had left off the summer before.

On May 11, the nomads set up camp on Flotten Lake, near the outskirts of the middle of nowhere, just 45 miles north of Meadow Lake, Saskatchewan, a frontier town with one-story frame buildings and muddy, unpaved streets. Accommodations were rugged, but the Allens were well adapted to living in "the boosh," as the locals called it. A Meadow Lake schoolmarm was horrified; how would the children learn their lessons? Mrs. Allen was bemused; there was more to education than stuffy classrooms.

Allen connected with pilot Bob Smith on June 6. They judged the breeding grounds to be farther north than previously searched areas; two 19th-century nest records placed cranes near Fort Resolution and on the Salt River in the Northwest Territories. The land's remoteness would certainly have protected whooping cranes from the harassment that plagued more southerly parts of their breeding distribution. Allen and Smith turned north, flying at low altitude on a path 5,760 miles (9,270 km) long across northern Alberta, north to Lake Claire, and finally to Great Slave Lake, where English explorer Samuel Hearne had trodden some 177 years before. Allen and Smith surveyed more than 23,000 square miles (60,000 square km) of crane habitat, turning back only when inhospitable tundra stretched endlessly before them. Allen's log read: "1:40 p.m. Rain squall. Rough going. Poor visibility all the way … this ends our search."

Allen returned home to find that the search for whooping crane breeding grounds faced mixed reviews. A prominent Canadian magazine had recently published an article describing the "bird's nest hunt that has cost $75,000 without an egg to show for it" and how wildlife experts planned to explore an additional 250,000 square miles (650,000 square km). The magazine was grossly mistaken. The search had not really cost the taxpayer anything; Smith was making these low-budget excursions as part of his regular duties and Allen was receiving his standard salary and limited expense money from the National Audubon Society. Regardless, misinformation of this nature would plague conservation efforts for years. *Life* magazine also printed a short article in 1947 that mentioned Allen's search as he trailed cranes north to their nesting site. Allen was pleased with the publicity that the article generated; unfortunately, some readers got the wrong idea. Many thought that Allen flew alongside them wing-to-wing at 40 miles per hour (65 kph), roosting where they rested, flying when they flew. Others suggested that the cranes should be affixed with radar bands that could be easily tracked — the Air Force was not amused. But how quickly a fool's folly becomes reality, given enough time. Nonetheless,

Allen spent many days answering letters, telegrams and phone calls, sorting out the interested public. Only one large section of potential whooping crane habitat was left to be searched; he made plans for the following summer.

Meanwhile, arrangements to unite Pete and Josephine were still underway. Allen considered the Aransas natural environment to be the best option but he was having some difficulty wringing money from the Fish and Wildlife Service to build a suitable enclosure. Furthermore, John Baker had recently met with George Douglass, director of the New Orleans zoo, to discuss moving Pete there as a temporary measure. Of course, Douglass was eager to cooperate. Allen remained skeptical; zoo diet, crowds, noise and city air would not be conducive to maintaining healthy cranes, let alone getting them to breed. He insisted that Douglass provide the birds with a cage sufficiently large for dancing, with ample fresh water and food, in a secluded area away from the gawking public. Baker reminded Allen of Douglass's obstinate nature: "For Pete's sake, don't tell him that his zoo isn't good enough!"

Douglass agreed to the conditions and acknowledged that if efforts in New Orleans failed, the birds would both be transferred to Aransas. He eagerly awaited Pete's arrival, scheduled for December 15. Early in February, Douglass reported that Pete and Josephine were getting along fine, but Allen decided to check for himself. He visited the Audubon Park Zoo on March 17, 1948, and was shocked at what he saw. Pete and Josephine were housed in deplorable conditions. Their small pens had been relegated to a filthy, noisy area near the zoo garbage heap, and they drank dirty water from old washtubs surrounded by muddy rat tracks. They had no shelter from the hot Louisiana sun except for a battered wooden shack at one end of the pen. Moreover, their zoo diet would never have promoted breeding; their daily food supply included only 1.5 pounds (675 g) of scratch grain, 1.5 pounds of chopped fish and 1.5 pounds of store-bought bread. Allen thought the cranes looked terrible; their color was poor and their bodies lacked tone. In Nebraska, Pete had been living in a large fenced area traversed by small river channels where he had freely caught fish and frogs. He was hysterical in his tiny zoo cage. Later, Allen would report his visit to the service in Washington.

> *Their present condition is a disgraceful thing! The species has suffered so much but those two birds represent the lowest rung. I could have wept when I saw them, if I hadn't been so numb with anger and shame.*

Allen recommended that the birds be taken immediately to Aransas and released on an isolated 150-acre tract of good freshwater nesting environment, well stocked with aquatic food, that was far removed from the Intracoastal Waterway but close enough to refuge headquarters for observation. He just needed $2,000 to purchase a fence. Certainly, Douglass would surrender the birds once he saw the rich life offered at Aransas. Allen's words were compelling and the fence was ordered. Washington arranged to get the cranes moved. Baker reassured Douglass that he would retain "title" to his bird; Douglass reluctantly agreed provided that they be returned to him if the "Aransas experiment" was a failure. Allen was decidedly relieved.

That winter there were 31 whooping cranes — including six juveniles — at Aransas, the largest number since James Stevenson had started counting a decade before. The wintering flock included a pair, which Allen called the Summer Pair, that had been injured by hunters during earlier migrations and had stayed on the refuge the previous summer. The largest, presumably the male, was known as Crip. Crip was a strong and stately bird, despite his badly broken wing. During the summer, he and his mate had leisurely fed together on 400 acres (161 hectares) of choice salt flats that was the Middle Pond Family territory during winter. When the owners returned from the north in October, they were not pleased with someone else occupying their prime real estate. The turf war that ensued ultimately ousted Crip and his mate to a poorer, unoccupied territory near Rattlesnake Point, but Allen was intrigued by the ritualistic calls and postures that comprised the battle. He decided to scatter corn near a blind to attract the birds for closer observations. All but the Middle Pond Family soon tired of the corn and went back to their fishing. However, when these resident cranes were tested by the neighboring Dike Pair, Allen was there with his field notebook.

The altercation began when the challenger issued a loud repeated *ker-loo! ker-lee-loo!* Middle Pond male stood menacingly tall, pointed his bill skyward and released an overwhelming whoop. His mate joined the chorus, and then suddenly the pair and their offspring flew in a low, skimming path directly into the Dike Pair's territory, where they arrogantly stood ground. The challenge was withdrawn and all parties resumed feeding on their own turf, boundaries intact, albeit with a few ruffled feathers. Both pairs faced an unusual problem later that season when a widowed female arrived at Aransas with her youngster. Her mate had been lost during migration, and Allen surmised that she had remained near his dead or injured body for some time, thus accounting for her tardy arrival on the wintering grounds. This "Extra Family" insisted on squeezing in between Middle Pond Family and Dike Pair, rather then settling in an unoccupied space. Middle Pond male was entirely intolerant of this female within the boundaries of his domain and would repeatedly bear down on her while on the wing, forcing her into the air, and chasing her mercilessly, bill to tail, often for an hour or more. Extra Family eventually retreated to wet swales beyond the live oak brush and peace resumed. Allen concluded that male whooping cranes were essential for territory establishment and maintenance. Without a mate, this female was able to do neither.

All was not well elsewhere on the refuge. Poachers killed Crip's injured mate in March. Allen retrieved the dying bird from the salt marsh, hoping that a veterinarian could save her, but she only lived a few hours. He examined her lifeless body; in addition to the shattered wing that had grounded her at Aransas, she had a recent bullet wound through the trachea. Her death was a focal point of Allen's annual report to Washington, in which he argued that Aransas could no longer provide whooping cranes with basic protection. He detailed other incidents: shrimpers taking target practice from their boats, hunters building blinds in crane territories, empty shotgun shells strewn around the reserve. Washington agreed to take Allen's recommendations under advisement.

The desperate hunt for the mysterious breeding grounds continued in June of 1948. Allen and Smith turned northward once again, this time painstakingly searching 42,000 square miles (109,000 square km) from Saskatchewan to Point Barrow, Alaska, and the Arctic Ocean shores. They found nothing, and Allen had run out of places to look. In August, he was forced to relinquish the formal search. In future years, the responsibility would fall to wildlife workers merely scanning the horizon for the great white birds on their regular rounds.

That autumn, only 25 of 28 birds that had traveled north in spring had returned, bringing with them only three young. The population was flatlining again. But high spirits were restored at Aransas when Pete and Josephine arrived in October. They were released into their large, fenced enclosure that included both brackish marsh, preferred winter digs, and freshwater cattail marsh that re-sembled traditional nesting habitat. Despite their long stay in captivity, the pair immediately became cranes. They stalked blue crabs, dug for worms and fished for shellfish and frogs. In December, Pete and Josephine performed a prenup-tial dance, abundant with bowing and whirling high leaps, and wedding invi-tations went out to John Baker in New York and fish and wildlife workers in Canada and the United States. The dancing continued frenetically until April 27, when all went quiet. When Bob Keefer approached the cattails, Pete drove him off, but not before he had seen the single egg in their nest. Four days later, the nest contained two eggs.

Allen set up an observation tower where he could watch the pair using a 19.5× spotting scope. He recorded their incubation patterns, something that had never been done before. He determined that the birds shared incubation duties and frequently changed places, as often as every hour or so during warm weather. Pete spent more time on the eggs during the day, but Josephine incu-bated through the night while Pete stood guard nearby. His duties included chasing off other birds, particularly egrets and herons, and killing and eating snakes. During the day, Pete would challenge intruders by whooping loudly, strutting his stiff-legged parade walk, and chasing the interloper until it fled. Allen also noticed that whenever the cranes left or approached the nest, they always did so through a jumbled, circuitous route through the cattails that would, hopefully, thwart the intentions of predators. Once the relieved bird was far enough from the nest, it would dance briefly or bathe with exuberance as if joyfully expressing its release from duty.

After 24 days, the birds stopped incubating. Allen found two broken shells in the nest; the eggs had been infertile. Although disappointing, this was not unusual. Early breeding attempts frequently fail. Mating must be precisely timed to produce fertile eggs, and birds use a variety of vocal and visual displays to coordinate their reproductive cycle. New pairs must learn their mate's ways before nesting success becomes routine. Unfortunately, Pete and Josephine never got a second chance. Two months later, the new refuge manager, Julian Howard, was awoken at dawn by Josephine's loud whoops of distress. Beside her lay her mate's body. A postmortem indicated death by natural causes, perhaps attributable to old age. Pete had been in captivity 12 years. Although no chicks had resulted from her first paring, Josephine had demonstrated a distinct willingness to breed. Therefore, after Pete's death, Allen was given the green light to capture Crip — the flightless male widowed the previous March — and place him with Josephine.

That autumn, all 30 migrants returned to Aransas, bringing four chicks with them. Allen was hoping that his public relations work was finally reducing the number of cranes killed during migration. That year, only one survivor remained from the White Lake, Louisiana, population. Unable to secure a mate, the lone crane would represent a considerable loss of genetic variation if it died.

Although they were legally protected by the Migratory Bird Treaty Act, whooping cranes continued to die by the gun, particularly during sandhill crane hunting season.

The decision was made to capture it and move it to Aransas. This bird, however, was not flightless. A risky plan was devised to hover a helicopter low and use the rotor's downdraft to pin the bird helplessly to the ground. Clarence Cottam, now assistant director of Fish and Wildlife, warned Allen and his team: "Be careful! Biologists are expendable; whooping cranes are not." The bird was eventually captured, but not without a long, arduous and stressful pursuit. After days of observation, Howard decided to feed the bird and release it near a freshwater marsh. When it wandered into another pair's territory, however, it was viciously attacked. Wounds were mended but the bird was still reluctant to eat. Nonetheless it was released again, this time far from other cranes. Howard saw the bird several times throughout the spring and summer and was encouraged that it might survive. However, on September 1, its dead body was found by the lake. Cause of death was undetermined but the autopsy indicated that Mac, named after the helicopter pilot, was a female. Josephine was now the only survivor of the White Lake genetic lineage; it was critical that she produce young. The incident left Bob Allen in sorrow, angered at their obvious collective ignorance in handling wild cranes. He would now vehemently oppose any future efforts to capture wild cranes.

In April, Josephine and her new mate, Crip, built a nest. When Josephine laid an egg, congratulatory letters, phone calls and telegrams again flooded into Aransas. Tourists came from as far away as New Zealand to visit the refuge, so many that Howard had to close the roads. Back in New Orleans, George Douglass was not too happy about the positive attention enjoyed by nesting birds and their "keepers." An article appeared in the *New Orleans Times-Picayune* on April 30.

> [Douglass] is anxious to get Josephine ... home. He plans to swap her,
> on a loan basis, to a northern zoo for a panda: 'The folks there have
> never seen a whooping cranes and our people haven't seen a panda.
> I expect our crane back after the egg is hatched so we can make the swap.

The chick, affectionately named Rusty, hatched on May 26, 1950. He was the first whooping crane to be born in captivity, and the happy news traveled to millions of interested admirers. A Corpus Christi newspaper article about the little crane sported a headline eight columns wide, dwarfing a royal birth announcement that appeared somewhere deeper in the edition. Rusty was healthy and frisky, and he trotted happily behind his doting parents as they foraged. Periodically, Crip would bend his long neck to the ground and gently offer his chick a small bit of food with the tip of his bill. When Rusty was four days old, Allen became distressed; he could no longer see him moving through the grass. When Allen and Howard entered the pen to investigate, Crip and Josephine looked up briefly, then bent their heads to the ground and resumed feeding. They made no effort to defend their chick. Allen's worst fears were soon realized. The two men searched the pen and found nothing but raccoon tracks. How Rusty died was never determined.

The black year of 1950 did not relent so soon. In October, only 26 adults returned from the north with five young; seven adult cranes had died since spring. Moreover, their safety did not improve on the wintering grounds. In January 1951, Allen saw people shooting at them from canal boats on several occasions. One juvenile disappeared and the body was never found. Allen had also turned back fisherman landing illegally on shore, and had caught two boys with a dead goose and a gun driving the refuge roads.

Drought conditions continued through the winter. Water levels remained low, but preparations for the breeding season began. Crip and Josephine were provided with a finer-mesh pen and raccoons were removed from the area; unfortunately, no one could do anything about the weather. On May 12, Josephine laid an egg in her nest, built on a small peninsula that extended into a tidal slough. All seemed well until a storm blew in. As the rising tide began to encroach on the nest, the panicking parents furiously attempted to build up its sides against the flood. Howard and his men intervened, gently placing the nest atop a bale of hay. Although the nest remained dry, the egg was later found smashed among the reeds. It had probably fallen out while the parents rearranged it.

That autumn, three more migratory cranes succumbed when their wings, legs and feet were blown to bits by gunshot. Adding insult to injury, George Douglass got tired of waiting and arrived unannounced at Aransas in December in the zoo's panel truck, declaring, "I've come to get my bird." The Fish and Wildlife Service was caught off guard. Howard called Cottam in Washington, who replied to keep Douglass waiting while he contacted the department's lawyers. But the appeal fell into the lap of a lawyer who did not care about whooping cranes, who decided that Josephine should be surrendered to Douglass. Cottam took the plea higher but received the same uninformed answer. Howard was forced to let Josephine go. Cottam was reluctant to split up the pair on the off chance that they might breed in the zoo. He offered Crip to Douglass, after he signed a release form acknowledging the government's ownership of the crane. Douglass triumphantly took the only mated pair of captive whooping cranes back to New Orleans and placed both on public display in a cage 10 feet (3 m) square.

When Bob Allen published his monograph on whooping cranes in 1952, it was apparent that the species' future had not improved considerably in several decades, despite the efforts of all involved. Aransas was no longer an isolated winter retreat; it hosted 20 oil wells, 50 regular oilmen and cattlemen, continuous road traffic and thousands of annual visitors. Not all were welcome, however. Poachers, squatters and hunters were continually violating refuge rules; in 1951, three men shot a doe at the headquarters' front door. Moreover, the whooper world population was dropping steadily. Only 19 adults and 2 young returned to Aransas in autumn 1952. At least eight birds were lost during migration, totaling 24 known deaths in two and a half years — more than the entire species population at that time. *Audubon* magazine reflected the concerns of Allen and others: "Each year it becomes more apparent that illegal hunting is a major factor in reducing the numbers of the whooping cranes." As before, a migration route–wide plea — posters, advertisements, radio broadcasts, newspaper and magazine articles, school programs — to stop illegal hunting was initiated. The National Audubon Society distributed a leaflet imploring shooters to lay down their guns, reading:

> *We appeal to the sportsmanship and humanity of every person, from Saskatchewan to Texas, to withhold fire, and to give any large white bird god-speed, instead of a charge of shot.*

Not everyone was understanding. Hate mail poured into service and society offices. Allen recalls a horrific solution to the "whooper problem" offered by a group of Saskatchewan farmers. They proposed that the best way to put the whoopla to rest would be to shoot them all out of the sky and be done with it, saving taxpayers both money and further concern.

The following year, John Baker attempted to convince George Douglass to relocate Crip and Josephine to a natural, enriched environment to encourage breeding. Their cramped new quarters behind the elephant house were only a slight improvement over the public exhibit cage. Baker suggested that the National Audubon Society Rainey Sanctuary, a large, semiwild and secluded tract of land in southwestern Louisiana where whooping cranes once lived, would suit them perfectly. Douglass replied that he was most concerned about the well-being of his precious birds and felt that the Rainey Sanctuary would not supply them with appropriate food sources. He bid Baker a fond goodbye and happy New Year.

Crip and Josephine had not nested since 1951 at Aransas. By 1955, all parties were concerned about lost opportunity; the cranes were not getting any younger. In May, Baker made his most recent of many requests to Douglass to consider alternatives before it was too late. Douglass reported that Josephine had just laid an egg; he did not have to consider anything. When the first egg was broken, Douglass eagerly awaited a second, which Josephine dutifully provided. Eager to cinch his ownership of the only captive breeding cranes, Douglass invited reporters and photographers to record the blessed event. Unfortunately, the preoccupied cranes were not cooperating photographically. When a visitor poked a stick into the pen to encourage them, Crip exploded in rage; while defending his mate and unborn offspring, he accidentally stepped on the egg and smashed it. Josephine would not lay another that year.

Tides finally turned in 1954. The hunt for the northern breeding grounds had been going on vicariously since 1949; Bob Smith and his colleagues kept watch as they went about their appointed tasks, but there had been no large-scale searches. Skeptics had their claws out: how hard was it to find a bunch of big white birds? In the eleventh hour, it happened. Twenty-four cranes left Aransas in April to places unknown. Three months later, Bob Allen received the long-awaited words. The telegram read simply:

> We have just received a telegram from WA Fuller, our mammologist at Fort Smith NWT, stating that four or possibly six whooping cranes were seen from a helicopter in Wood Buffalo Park, NWT, on June 30th. The group included a pair with one young.

They had been found by helicopter pilot Don Landells and forester M. G. Wilson, who were flying between Fort Smith and a forest fire near Hay River. Allen and Smith had traversed that corner of Wood Buffalo Park seven years earlier; however, at the time, the pristine, tamarack-dotted muskeg had been obscured by heavy rain. Allen was overjoyed with the news, even more so when he realized that the cranes were nesting safely within the boundaries of a remote national park.

Wood Buffalo National Park is the largest in Canada, covering more than 11 million acres (4.5 million hectares) in northeastern Alberta and southwestern Northwest Territories. Established in 1922 to protect woodland bison, the park is a natural paradise of shallow lakes and ponds, mixed brush and prairie, waterlogged muskeg, stunted trees and timber-choked rivers. It was virtually impenetrable in summer and was typically avoided by aircraft because it offered no emergency landing sites. So, it remained unexplored for years. Although Allen was anxious to see nesting cranes, he agreed it was too late in the season to launch an expedition. Plans were made for the following year. Canadian authorities promised to keep an eye out, enforce a ban on unauthorized aircraft and keep the location a secret to discourage trophy hunters. This last promise was not gratuitous. The young crane that had been seen beside its parents in July, confirming the existence of northern breeding grounds, never made it to Aransas that fall. No young survived the trip south in 1954, and three adults were lost as well. Howard's annual report to Washington lamented: "With a drop of three, and no young, the situation becomes ever more critical."

Bob Allen and his companions, including Ray Stewart of the Canadian Wildlife Service, set out to explore the breeding grounds in May 1955. Stewart had recently flown over the area, noting the location of white dots that were nesting whooping cranes on the swampy islands. The ground search was not so easy. Beset with deplorably rough terrain, unnavigable rivers, lost directions and insatiable mosquitoes, the searchers took weeks to reach the nest site. Eventually, Allen was able to enter in his notes: "It has taken 31 days and a lot of grief, but let it be known that at 2 p.m. on this day 23rd of June, we are on the ground with whooping cranes! We have finally made it!"

They explored the whooping cranes' other world for 10 days, and determined that breeding grounds conditions were generally ideal. Food, water, space and isolation were plentiful, and predators were scarce. In addition, fire was little threat in the soggy habitat. They found at least 11 adults with six young (including two sets of twins) between the Sass and Nyarling rivers, and two additional pairs and some nonbreeding adults near the Klewi River. No doubt the entire breeding population nested in this secure, remote area. Everything was looking up, for now.

While Allen was searching northern breeding grounds, the news came into Aransas that the Air Force was planning to use Matagorda Island for photoflash bomb testing. Photoflash bombs had been designed to emit a brief blaze of light with the intensity of 700 million candela (candlepower), the equivalent of 100,000 100-watt bulbs. They had been used during World War II for medium-altitude night photography or to illuminate bomb targets. Photoflash bombs were a big problem for wildlife. When they were tested previously in Oklahoma, local waterfowl left the area and were long in returning. This could spell the end of the Aransas flock if they were forced off the refuge for any period of time; the Louisiana hurricane of 1940 still lingered in memory. Protests were launched by oil companies, landholders, hunters and fishermen, none of whom wanted any restrictions on their activities. The National Audubon Society and the Fish and Wildlife Service spoke on behalf of the cranes. The loud rallying cries made little difference and the Air Force decided to go ahead with their plans regardless. But the cavalry arrived just in the nick of time, and it was brandishing maple leaves. Since the discovery of the northern breeding grounds in Wood Buffalo National Park, the whooping cranes had become Canadians. On October 20, 1955, the Canadian Embassy in Washington dispatched a note to John Foster Dulles of the US State Department. It read:

> *The Canadian Ambassador presents his compliments to the Secretary of State of the United States of America and has the honor to draw his attention to a proposed photoflash bombing range … It is hoped that the existence of the whooping crane will not be imperiled at a time when the prospects for the increase of the species have at last become bright.*

The reply from the State Department was received 10 days later.

> *The Secretary of State presents his compliments to His Excellency the Ambassador of Canada and has the honor to acknowledge the receipt of his note No. 657 of October 20, 1955, drawing attention to reports of a proposed Air Force photoflash bombing range. The Department of State*

informed the Department of Defense of the Canadian concern, and has now received a reply stating that the Department of the Air Force is withdrawing its proposal for extension of the Matagorda Island Air Force Range, which is the matter referred to in the Ambassador's note.

The first wintering cranes arrived in relative peace on October 18; by November 4, parents and twins had landed. That year broke the whooping crane's losing streak — the number topped 28, the highest since records had been kept at Aransas.

George Douglass played godfather again in the spring of 1956 when Josephine — now at least 16 years old — laid an egg on the bare ground of her tiny enclosure. A second egg was laid four days later. Douglass, always ready for publicity, hailed the newspapers and photographers but kept them out of sight on the elephant house roof. Although both eggs hatched successfully, the young did not survive long. Two days after hatching, the younger chick was taken by a predator, perhaps a rat that could have easily passed through the cage bars. The older chick lived 45 days before it was found collapsed in the well-trampled grass. An autopsy indicated that it had died of aspergillosis, a respiratory fungus that afflicts domestic poultry. The zoo had no veterinarian on staff.

After the chicks' death, interested parties again raised concerns about the entire captive breeding stock, two birds, being managed by a zoo that obviously lacked the competence to ensure success. Should Crip and Josephine be taken from George Douglass by court order? Should a second captive breeding program be started elsewhere? The concerns deepened when Rosie showed up at the San Antonio Zoo.

Rosie had left Aransas, heading north, in May 1956 but was grounded with a broken wing when she flew into a power line near Lampasas, Texas. She sought refuge beside a ranch house where an open water tank offered respite from the blistering heat. The ranch owner threw a lariat around her neck and dragged her into his house. The call came in to Fred Stark, director of the first-rate San Antonio Zoo — a rancher had a big white bird tied up in his kitchen. Unlike Douglass, Stark did not claim Rosie as property. He contacted Fish and Wildlife, which, after deliberation, decided that Rosie would remain at the zoo on loan from the service. She was provided with luxury accommodations, befitting her status, and awaited a mate. It would take nine years.

It was at this time that the vision of whooping cranes' future became murky, with opposing camps divided down two lines of belief. One side — whose members included Yale University professor Dr. S. Dillon Ripley, Fish and Wildlife biologist John Lynch, and Saskatchewan's Fred Bard — generally maintained that the species was doomed to extinction, and that intensive aviculture was their only salvation. Ripley went so far as to suggest that the entire Aransas flock should be captured and pinioned (rendered flightless by cutting off their wing tips) to keep them grounded in Texas. Rosie became the figurehead for their cause.

Bob Allen, on the other hand, was gravely opposed to taking birds from the wild, or grounding those that still clung to their ancient legacy of migration. Moreover, he doubted the viability of ever successfully returning captive-bred whooping cranes or their young to the wild. He did not want the species

Devoted mother Josephine tends to her two newly hatched chicks at the Audubon Park Zoo in New Orleans.

to become a flock of mere zoo specimens. John Baker was also opposed to capturing wild birds. He suggested that the proposal was based entirely on the erroneous assumption that the species was doomed, and added that if zoos wanted cranes, their source should be birds already in captivity. They cited ill-fated attempts to capture wild cranes in the past; for every crane captured, others would certainly be lost or injured. It was not worth the risk. A party of 40 interested men met on October 29, 1957, to discuss the issue. Members on both sides would not be moved. Moreover, the National Audubon Society, represented by Allen and Baker, found itself with the curious task of convincing aviculturalists that the wild crane flock was adequately robust to survive without intervention — this after decades of pleading for support to bolster the declining species. No decisions were made, although a new committee was struck to study the problem. To the dismay of Allen and Baker, the new Whooping Crane Advisory Group, comprising many "spectators-turned-advisors," would supersede the idealistic Cooperative Whooping Crane Project that had served so well for 11 years.

Meanwhile, George Douglass was still staving off attempts to relieve him of his cranes. The Fish and Wildlife Service decided that it might be

Seemingly despondent parents Josephine and Crip pace in their tiny zoo enclosure following the death of their second chick from a fungal disease at age 45 days.

more time-efficient to send expertise to New Orleans rather than move Crip and Josephine. Ripley and Fairfield Osborn, the New York Zoological Society president, arranged to pay the salary of retired bird keeper George Scott; Ripley and Dr. Frederick Lincoln, representing the Whooping Crane Advisory Group, went to New Orleans to woo Douglass into accepting their assistance. The deal was finally struck as the three men sat in the crane pen; while Crip and Josephine circled them, yellow eyes staring intently. The tricky issue of ownership of the cranes' progeny was raised upon their return. Lincoln was reluctant to face Douglass again. He maintained that since whooping cranes are migratory birds receiving all benefits of the Migratory Bird Treaty Act, they cannot be owned by anyone without due permit and process. He considered the question purely academic, and thus the answer need not be determined until the occasion rose.

The occasion arose later that year, when Josephine laid two eggs. George Scott was in New Orleans to supervise, and both chicks survived under his care. They were named George and Georgette, in honor of Douglass and Scott. Lincoln recommended that Crip and Josephine stay in the Audubon Park Zoo now that they were successfully producing young, but their offspring would be moved elsewhere the following spring. But Douglass was blissfully flaunting his brag rights as the world's only whooping crane breeder. At a National Audubon Society meeting, he spoke on "The Reproduction and Rehabilitation of the Whooping Crane" and passed molted feathers around the room as he explained how juveniles changed from cinnamon to white. In November, the United States Post Office issued a commemorative stamp depicting the birds. Douglass had many activities planned in New Orleans that day, including a television press conference with two senators, several congressmen and the crane family, which had been removed from their pens for photo opportunities. Although the program proceeded without incident, it would be a long time before both Douglass and the Audubon Park Zoo would recover from criticism that befell them that day.

At the San Antonio Zoo, Rosie was still awaiting a mate. That summer, a single bird remained at Aransas after the flock had flown north, and efforts were made to capture it. It sustained no injuries but died suddenly anyway. This event has since been attributed to capture myopathy, caused by overstimulation of the nervous system during extreme stress. Baker readily used this incident to squash notions of taking more cranes from the wild. Rosie would have to wait for a bird to be wounded accidentally. Reports of two such casualties came in from Oklahoma and Manitoba, but word was not received on time. Neither bird found its way to Aransas that autumn.

Fish and Wildlife, renamed Bureau of Sport Fisheries and Wildlife, was still having crane problems in New Orleans in the winter of 1957. Lincoln had given Douglass a permit to keep Crip and Josephine with an appended proviso that the bureau "reserved the right to take them into custody at any time."

He reminded Douglass that their young were also government property and that they should be moved to another facility, such as the San Antonio Zoo, to avoid a catastrophic event obliterating the entire captive population. Douglass replied that his "curator of birds" was studying their plumage development and they could not be moved in the near future. One need not look far to spell disaster at the Audubon Park Zoo under Douglass's direction.

During the next two breeding seasons, Josephine laid ten eggs but only a few were fertile. Despite Scott's efforts, only one chick survived in the end. He was christened Peewee. Everyone hoped that these failures were just bumps in the road of raising whooping cranes and not the shape of things to come. After all, the birds were of undetermined age; eventually, their reproductive life would be over. Daniel Janzen, director of the Bureau of Sport Fisheries and Wildlife, asked the solicitor for the Department of the Interior to "look into legal possibilities" for removing the cranes from the New Orleans zoo. The answer was returned three months later: the government had neither title to the adults nor any claim to their offspring. The bureau would have to resort to pleasantries, not court orders, if they wanted to get anything out of Douglass.

While the bureau was battling over the captive cranes, things were not improving for the wild birds. Allen had always been troubled by the prospect of Aransas not being large enough to host all cranes on quality territories should the flock grow considerably in size. A proposal to increase the refuge to include areas where cranes typically wintered beyond current boundaries, and to rename the refuge the Whooping Crane National Wildlife Sanctuary, was put forth to congress. It was defeated late in 1957. Allen returned to Aransas after a long illness and was disappointed that the state of affairs had not changed in his absence, despite his detailed recommendations. Fortunately, wild cranes were looking after themselves, increasing slowly to 32 after producing a bumper crop of nine young in 1958. Allen made another attempt to increase the refuge's size. On his urging, the National Audubon Society convinced local landowner Toddie Lee Wynne to hold lease on an additional 5,720 acres (2,300 hectares) on Matagorda Island for the cranes' use.

Bob Allen retired from the Audubon Society in June of 1960 at age 55, after 30 years of dedication to the cause of endangered and threatened birds. He continued to march for the cranes even into his retirement, but over the years their fate had been gradually been taken out of the hands of any one man — even if that man was Robert Porter Allen. Yet Allen steadfastly believed in whooping cranes' ability to survive, but to do so only on their own terms, although there seemed little cause for hope in those days. In Allen's own words,

> *As the human population curve goes up, the Whooping Crane curve goes down. This is a bird that cannot compromise or adjust its way of life to ours. Could not by its very nature: could not even if we had allowed it*

The whooping crane's remote northern breeding grounds in Wood Buffalo National Park were finally discovered in 1954 after almost a decade of desperate searches.

the opportunity, which we did not … If we succeed in preserving the wild remnant that still survives it will be no credit to us; the glory will rest on this bird whose stubborn vigor has kept it alive in the face of increasing and seemingly hopeless odds.

Allen had promised himself that upon retirement he would devote his time to writing. He moved back to Tavernier, Florida, and finished writing *Birds of the Caribbean*. He was working on additional studies of flamingos and roseate spoonbills when he died of a heart attack on June 28, 1963. When Bob Allen passed, whooping cranes lost their greatest ally; they would not find another of such dedication for a decade.

The Bureau of Sport Fisheries and Wildlife finally made a significant step in whooping crane recovery late in 1960. They gave up dancing with George Douglass and started their own captive propagation program. The project was given to bureau biologist Dr. Ray C. Erickson. Erickson acknowledged that no one really knew much about breeding whooping cranes in captivity. Douglass's lack of success over the years was a testament to that. He suggested that it would be more sensible to use their abundant relatives — sandhill cranes — to research techniques of capture, captive propagation and reintroduction into the wild. Wise thoughts, but unfortunately, it would take five years for Erickson's proposal to come to fruition.

In the meantime, Erickson searched for reasons to explain, specifically, why the wild flock was increasing so slowly in size. They were producing young but, somehow, most immature birds were never recruited into the breeding population. Many were not living long enough. Earlier studies had suggested that nonbreeding adults — typically one to four years of age — were summering somewhere other than Wood Buffalo National Park. Erickson also noted that in years where a great many nonbreeding adults were lost, there were often many young produced. The problem could not be breeding-ground conditions. Moreover, the dark years were typically coincident with wet prairie weather. Erickson determined that when water levels were high in Saskatchewan and potholes and sloughs were full, nonbreeding cranes did not need to fly to the Northwest Territories to find satisfactory summering habitat in abundance. These conclusions were supported by increased whooping crane sightings southeast of Regina in wet years. Lingering in the south, however, these birds faced the wrath of hunters and farmers as they foraged in adjacent croplands. Exacerbating this problem were 1959 changes to provincial legislation that allowed landowners to shoot sandhill cranes on private property if they were deemed to be agricultural pests. Erickson was not in favor of capturing wild whooping cranes to feed a captive breeding program. However, he was able to recommend that if absolutely necessary to take birds from the wild, then a few immature whoopers could be captured for breeding stock in Aransas when a wet prairie summer was predicted — based on groundwater levels and snowfall depth — without having much impact on the wild flock. Nonetheless, he urged the bureau to begin sandhill crane experimental programs first.

During the early 1960s, the wild flock was facing hazards on all sides. In the north, the Canadian National Railway proposed a new rail line through Wood Buffalo National Park. Prime Minister John Diefenbaker, mindful of public protests, requested extra funds from Parliament to survey a western route around the park. Fortunately these funds were secured, and the alternate route adopted at substantial cost. Later that year Hurricane Carla, the eighth-strongest storm in US history, made landfall at Port O'Connor, just 25 miles (40 km) from Aransas. Carla pounded Matagorda Island with sustained winds of 115 miles per hour (185 kmh) and a storm surge 10 to 18 feet (3 to 5.5 m) above normal tide levels. When it was all over, Carla had left her mark on crane wintering grounds, but the jury was out as to whether scars that remained served good or ill from the bird's perspective. The hurricane had caused extensive damage to houses, businesses, boats and piers that had been encroaching insidiously on the refuge for years; it wiped the slate clean, so to speak. However, Carla's storm surge had carved several washover channels through the low-lying barrier island and silted up many prime feeding areas on Matagorda and Blackjack Peninsula.

With blue crab populations declining, refuge workers began setting out grain as a supplemental food source for the returning cranes, and artificial feeding continued through the 1960s when natural food was scarce. Refuge workers learned a valuable lesson. They found that they could reliably lure cranes away from potential danger by providing food, particularly useful in the event of an oil or chemical spill on the Intracoastal Waterway. The strategy also paid off when Air Force activity increased on Matagorda Island in 1962. Four cranes were lost during the Cuban missile crisis; others were pushed back into poor-quality habitats. Artificial food gave these cranes feeding options. Precisely what happened to the missing birds was never determined because no carcasses were found. This was by no means unusual; a Texas ranch foreman had once admitted that he'd "sooner be caught with an illegal whiskey still than have anything to do with a dead whooping crane." All told, 16 adult cranes were lost in the first three years of the 1960s, and no young survived from summer 1962. By 1963, the Aransas population had dropped to 32 from a record high of 39 and, overall, had lost five years of growth. Aviculturists seized the opportunity to declare the species "needlessly doomed to extinction" and pushed to breed them in captivity.

Douglass's efforts in New Orleans, however, were not advancing the cause. Josephine had laid 16 eggs between 1959 and 1961 with only one live chick, named Pepper, resulting. Two others had died prematurely, and most of the remaining eggs were infertile. George and Georgette were four years old in 1961, and Peewee was three. Douglass, unaware of the evils of inbreeding depression, which causes poor health and lower levels of fertility, was anxiously hoping that any two of the younger birds would mate. By instinct, the siblings were much wiser than Douglass, and they fought each other with such animosity — as do many sisters and brothers — that they had to be separated regularly to reduce the bloodshed. The bureau made repeated "suggestions" to Douglass to offer one of them to the maiden Rosie at the San Antonio Zoo, but he just replied that there was little point because no one knew whether they were male or female. Like most birds, cranes do not have external genitalia to expedite the process. Not to be outdone again by Douglass, the bureau looked for a biologist who could determine the birds' gender by some other means.

That biologist was Roxie Laybourne, a pioneering forensic scientist at the Smithsonian who was better known for her remarkable ability to identify bird species by feather structure, a skill frequently utilized by the FBI and FAA to solve otherwise enigmatic cases. Laybourne and a colleague at Johns Hopkins University had just invented an instrument called a cloacascope that could be inserted into a bird's cloaca — the common body opening of most vertebrates, except mammals, that serves to release products of the excretory, digestive and reproductive systems — to visually examine its internal organs without physical damage to the bird. Perhaps this could be used to sex Douglass's cranes, if it could be demonstrated that it worked.

The previous year, the Bureau of Sport Fisheries and Wildlife had posted an open hunting season on sandhill cranes, causing considerable alarm among National Audubon Society members. If hunters were unable to tell a whooping crane from a goose, how could they possibly hunt sandhill cranes with discernment? Nevertheless, the hunting season had one advantage: it provided Laybourne with a multitude of dead cranes to practice on. Checkpoints were established where cranes were being shot, and hunters were required to temporarily surrender their kills while Laybourne took a peek. From there, she moved up to live sandhill cranes housed at the National Zoo in Washington and, along the way, developed restraint devices and handling protocols to reduce stress and injury to the birds during the procedure. In spring 1962, Laybourne packed up her cloacascope and avian straightjacket and traveled south to the Audubon Park Zoo to examine the four young cranes. It was no surprise to find that Douglass had gone out, leaving word that no one was to touch his cranes. Laybourne left New Orleans, continued west to San Antonio and verified that Rosie was a female.

He may not have expressed it to others, but Douglass's overconfidence was waning. His success as captive-crane keeper was far from stellar, and now his determination to maintain absolute control had prevented him from knowing his birds' breeding potential. He turned to an acquaintance at the Purina Chow Company — where he bought the cranes their Game Bird Crumbles and Purina Checkerettes — for advice. Purina operated an animal laboratory in St. Louis that was used primarily for research on poultry. Perhaps Douglass could use this facility if he ran into difficulty. He called the Purina man later that year when Josephine produced a chick with a malformed right leg from a slipped tendon. Douglass sent the 10-day old chick to St. Louis, where a veterinarian operated to correct it. When the stitches tore out, a second procedure took place, but the chick died on the operating table. Josephine's second clutch was sent to the Purina lab for incubation. The eggs were fertile, but they failed to hatch. Douglass decided to let the cranes incubate their third clutch. Only one egg hatched, but the chick died a week later from an undetermined cause. The second chick died while still in the shell.

Dillon Ripley was outraged, as were others invested in the whooping cranes' survival. He dispatched a biting letter to the Assistant Secretary of the Interior accusing the government of doing nothing to prevent the species' extinction. A polite reply from the Assistant Secretary's office reminded Ripley that, in the opinion of the solicitor, the cranes and their offspring belonged to the Audubon Park Zoo. Ripley's second letter accused Douglass of being a man who "for one reason or another, seems to have taken extraordinary pains to jeopardize the future of the species." The fates intervened and Douglass was finally coerced into surrendering Josephine's first clutch of 1963 to bureau aviculturalists in New Mexico.

Josephine laid an egg on March 12 but proceeded to destroy it and two subsequent eggs; they were all infertile. Her fourth egg was grabbed before she could put a bill to it, and sent to New Mexico. Josephine laid four more eggs that year, none of which produced chicks. An April *Life* magazine article titled "Whoopers' Bloopers: Cranekeeper Is on the Spot" featured a photograph of Douglass in commiseration over three broken crane eggs, as text described how 34 eggs had yielded only four surviving chicks. It further implied that his limited success had resulted from outside help in the guise of George Scott, which was ostensibly true. Douglass was decidedly furious, claiming that the bureau had engineered this bad publicity. The deal was off; they would get no more eggs. To make matters worse, the egg in New Mexico was infertile.

The Whooping Crane Conservation Association, however, still had a hand to play. This group of devotees, previously called the Whooper Club, had been actively involved since the 1950s in supporting legislation and restoration projects that promoted the survival of whooping cranes and their habitat. The president of the WCCA in 1963 was Jack Kiracoff, a breeding bird hobbyist. Kiracoff deviously planned to restore working relations with Douglass by playing to his delusions of grandeur in a public acknowledgment of his "success" in breeding rare birds. Despite the fact that the International Wild Waterfowl Association, then the parent organization of the WCCA, had recently condemned Douglass as a primary hindrance to whooping crane survival, Kiracoff suggested that it might be beneficial to present Douglass with the George Allen Memorial Award for his "outstanding contribution to aviculture" at the annual IWWA convention in October. WCCA members were justifiably aghast, but Kiracoff followed through with his plans, regardless. Of course, Douglass took the bait; he would be honored to accept the award. On presentation day, however, Douglass failed to show up. Convention delegates were already assembled in the hall when a message was received. Douglass had been detained in his motel room — someone had stolen his money and all his clothes, as well. He was unable to leave his room. An *ad hoc* apparel subcommittee was struck and dispatched *ex tempore* with tape measure in hand to suit him up more appropriately for the occasion. Kiracoff's plan worked. Douglass accepted the award in good faith, and renewed his offer to share Josephine's eggs with the bureau.

In the spring of 1964, the bureau made ready their facilities at the Department of the Interior Wildlife Research Substation at Lafayette, Louisiana. Government aviculturalists were anxious to try natural incubation and hatching methods in hope that they would achieve better temperature and humidity control; John Lynch had spent the winter conditioning Japanese Silkie bantam hens to act as surrogate mothers. Silkies are devoted setters but incubating an egg about half their size is a monumental task. They can be trained, however, to perch precariously atop the huge olive-brown eggs by systematically substituting larger and larger eggs for their own until they are fully comfortable with the idea. This procedure makes good use of an avian behavioral strategy by which parents differentially favor larger offspring. Larger tends to equal greater survivability, and the response is to lavish more attention on it. This is how cuckoos and other brood parasites — which lay their eggs in other bird species' nests, leaving child-rearing to foster parents — get away with it. By choosing small host parents, they guarantee that their much larger chick will be well tended. Hosts actually prefer the alien chick to their own: "Just look at this wonderful big chick I've made!" Very little else matters.

So two Silkie bantams — Patience and Petulance — began their vigil around Easter; seven backup hens waited in the wings. Replacements were not needed, however, since the A-team did not abandon their post. Nonetheless, the first trials did not end successfully. Bureau aviculturalists, and Patience and Petulance, failed to realize that whooper parents instinctively allow their eggs to cool at intervals during incubation to extend the time for embryological development. Consequently, the first egg hatched prematurely. The second hatched chick seemed more viable, although it had a defective leg muscle. It died of a hemorrhage 18 days later. Josephine laid eight more eggs that season, but they were all infertile. Everyone feared that either Josephine or Crip or maybe both birds were quickly approaching the end of their reproductive years. It was imperative that Rosie, who was considerably younger, join the ranks. Earlier, Douglass had offered to introduce Rosie to Crip and Josephine's offspring. The bureau finally agreed, but not before obtaining a legal agreement stating that her young would be federal government property.

Nineteen sixty-four was a watershed year for whooping cranes, despite failures in Lafayette. The wild flock brought an unprecedented ten young back to Aransas, and there were now eight cranes in captivity. The most recent contribution had been spotted on the nesting grounds by a Canadian Wildlife Service pilot. The juvenile, one of a pair of twins, was dragging a broken wing; obviously, it could not join its family on the upcoming trip south. Permission was granted to capture it. The first few weeks were tentative, but the young bird survived. He was christened Canus in recognition of Canada and the U.S. and their cooperative venture. Roxie Laybourne subsequently pronounced him a male.

Later that year, another chapter in the annals of whooping crane recovery closed: George Douglass died. The Associated Press touted him as the only man to breeding whooping cranes in captivity, but people closer to the story better understood the darker side of this honor. Douglass himself may have known this best of all; for each fleeting moment of success, there must certainly have been a grain of doubt.

William Pohlmann, a horse breeder, was named new director at the Audubon Park Zoo. Unlike Douglass, Pohlmann enthusiastically invited collaboration with the Bureau of Sport Fisheries and Wildlife. Laybourne visited soon after and determined that George, Georgette and Pepper were males — hence, their hostility to one another — but Peewee was a female. Pohlmann also accepted the bureau's offer to build a new crane facility at the zoo, and to provide a biologist to oversee future reproductive efforts. He agreed that offspring should be distributed elsewhere as a hedge against a stochastic total wipeout. The following year would yield no young, however. Rosie and George were getting along fine, but they did not build a nest. Peewee and Georgette (renamed George II), on the other hand, displayed their usual animosity to one another; they were siblings, after all. Josephine did her duty and laid three eggs that year. Two were removed to be incubated by surrogate hens, but proved to be infertile. The last chick died before it hatched. Josephine would never lay again.

In 1965, Hurricane Betsy made a direct hit on New Orleans, killing 75 people. Although the birds safely weathered the storm, its aftermath proved fatal. A helicopter hovered low over the zoo to survey the damage. Driven mad with fright, some cranes rushed around their cages; others cowered in terror. Josephine leapt wildly and flapped her pinioned wings, battering her hysterical body against the wire fence of her enclosure in a thwarted attempt to escape the noise and confusion. She died the next day. An era of captive crane breeding had ended. But, perhaps, another more hopeful one was just beginning — this time, without the ego of a self-proclaimed giant holding the reins.

The mid-1960s saw the dawn of conservation for species' sake, not because they had special recreational or commercial value to humans, but because they were worthy of salvation. Perhaps the foremost acknowledgment of this *volte-face* was the passing of the Endangered Species Preservation Act (forerunner of the Endangered Species Act) in 1966. Purportedly inspired by the whooping cranes' plight, this legislation provided legal means and encouragement for the protection of listed species and their habitat. Naturally, whooping cranes were among 36 bird species on the inaugural list. The Endangered Species Act was not perfect then — and it still is not now — being too often bogged down in red tape and invested interests. Yet it fostered a sense of enlightenment and consciousness among everyday people, and many North Americans began to develop responsibility that extended beyond the needs of *Homo sapiens*. When the list was finally posted, there were 44 whooping cranes alive in the wild, and 8 cranes in captivity — the highest number since the early days of the 20th century, but hardly a success. Although all involved appreciated this fading species' tenacity to cling to very shreds of existence, they knew that whooping cranes had lingered near the brink for too long — something had better break, and quickly.

Around this time, the ears of Washington were beginning to realize that average Americans were becoming more concerned about declining wildlife, and they responded with votes. And, as so many times before, another hero in the whooping crane story came out of the blue. In previous years, South Dakota Senator Karl E. Mundt, whose state office was on the whooping crane flyway, had supported funding to study sandhill cranes. Mundt was an avid outdoorsman who well knew the importance of experimental sandhill crane programs as precursors to whooping crane projects. When the Bureau of Sport Fisheries and Wildlife announced its desire to establish a research facility dedicated to reverse the extinction trend in native birds and mammals, Mundt made himself heard. He pushed through Congress an amendment to an appropriation bill that brought into being the Endangered Wildlife Research Program at Patuxent Wildlife Research Center, near Laurel, Maryland. The wildlife center had been in existence since 1936, but had previously been dedicated to captive propagation and the study of pathology (causes and effects of disease) in poultry, all very useful skills on which to base endangered species recovery programs. Mundt's amendment provided additional funds to hire staff and for construction of new avicultural facilities to house sandhill cranes in a pilot project for whooping crane captive breeding. Bureau biologist Ray Erickson, who had proposed a captive whooping project so many years ago, was made director of the new facility.

Among the wildlife center's primary goals was to establish captive populations of endangered animals to bolster wild numbers either by producing captive-bred individuals to be released or by producing young that could be fostered to wild parents. Captive breeding would also allow biologists an excellent opportunity to study the animals' physiology and behavior, and would serve as a useful platform for public education. Whooping cranes, California condors, black-footed ferrets and ivory-billed woodpeckers were among the first species to be short-listed for inclusion. Construction was completed in 1966 and the program got underway.

Meanwhile, Crip and Rosie — without the hindrance of George Douglass's private agenda — were moved to the San Antonio Zoo, where they quickly set up housekeeping. They successfully hatched two eggs the following spring. One chick died but the other, named Tex, survived. Tex's early grasp on life was tenuous, which necessitated hand rearing. Zoo director Fred Stark lavished all his attention on the little bird. The San Antonio Zoo had been around since 1914 but Stark had a vision and dedication to zookeeping that outreached his predecessors'. As the zoo's new director in 1934, Stark and his partner Richard Friedrich — credited by some as the real inventor of air conditioning

— converted an abandoned rock quarry into one of North America's first "cageless" zoos, where animals had space to roam and visitors could observe them in more natural conditions. Stark had been keen to participate in captive crane programs for decades, but was forced to wait in the wings while a lesser man stood in the spotlight. When Tex was born, Stark realized one of his dreams. He kept vigil over the chick almost day and night, feeding her mealworms by hand and bolstering her bones with calcium supplements. Tex, in turn, thrived; however, it was during these early days that she developed an unusual preference for dark-haired men, a predilection that would one day win America's heart on late-night television. During the summer of 1967, Stark transferred his teenage charge to Erickson, who took the chick to Patuxent. Tex had never seen another whooping crane and it was time for her to learn her true nature. Fred Stark died of a heart attack shortly after. He was greatly missed, and not the least by Tex.

Biologists were not without concern, however. For one thing, the entire program was based on the assumption that what worked for sandhill cranes would also work for whooping cranes. And although a captive propagation scheme can keep the species' genetics alive, the birds themselves become little more than a very large zoo exhibit if they cannot be placed back into the wild again, acting like real cranes. This has been the major stumbling block of most captive rearing programs. Captive animals lose useful behaviors as quickly as they gain bad habits. Moreover, their offspring know nothing about being wild, and frequently perish from starvation or predators soon after release. Improper imprinting on animal handlers also occurs epidemically, following which newly released birds fail to recognize visual and vocal behaviors associated with courtship and physical spacing that are presented by wild members of their own species; no communication equals no reproduction. Nonetheless, the program proceeded. These were all details that could be resolved later.

In May 1967, a party flew north to Wood Buffalo National Park to remove eggs from wild nests for transport south to Patuxent Wildlife Research Center. Among the crew were Ray Erickson and Canadian Wildlife Service biologist Ernie Kuyt. Kuyt had recently mapped the northern cranes' nest sites and calculated the size of their breeding territories. He would later initiate a color-banding and radio-telemetry program to track the cranes from Wood Buffalo to Aransas. In 1967, Kuyt and Erickson removed six olive-buff eggs from 17 that were observed in nine nests. They were successfully transported to the Patuxent facility, where five hatched. The following year, nine of ten eggs taken from the wild hatched and seven chicks were reared to six months of age. Egg removals in 1969, 1971 and 1974 yielded 34 more eggs to total 50, from which 37 hatched. Surveys showed that, thankfully, wild whooping crane production had not declined since the egg removal project began; in fact, population growth had increased slightly. Unfortunately, Patuxent could not make the same claim. As it turned out, raising whooping cranes in captivity was not as simple as following sandhill crane guidelines. Although biologists had perfected techniques for handling and incubating whooper eggs, it was not easy to raise healthy chicks to adulthood. Through the years, they struggled to overcome losses due to disease, trauma, leg malformations from a rich diet and lack of exercise and, perhaps most of all, insufficient funding. But by 1975, 21 adult cranes lived in pens on the 4,200 lush wooded acres (1,700 hectares) that was Patuxent. Such is the way of pioneers — every tragedy yields a little more enlightenment. The challenge now was to encourage these cranes to breed.

Nonetheless, Patuxent's captive crane propagation program had become the object of criticism in the early 1970s because some conservationists believed that removing eggs was taking future reproductive members from the wild flock. Although true in some degree — the potential to put captive-bred birds back into the wild flock had not yet been assessed — one could not deny that workers at Patuxent had raised 21 viable cranes in the same eight-year period that the wild flock had increased by only eight individuals. Mortality along the migration route was still disgracefully high; in 1972, for example, 59 birds left Aransas in spring but only 46 returned that autumn. They were still dying by the gun, particularly during sandhill crane hunting season. But researchers at Patuxent knew that their immediate goal was to make the captive flock self-sustaining so that eggs no longer needed to be taken from wild crane nests. Better still would be the production of extra individuals that could be released to bolster the existing Wood Buffalo–Aransas flock, or establish a new flock elsewhere.

However, getting captive birds to court and mate presented as many challenges as chick rearing, albeit different ones. For one thing, adult cranes are monomorphic: both sexes look exactly the same. Although males are typically larger than females, there is considerable overlap in size. Workers were reluctant to subject them to a cloacoscope examination. This procedure is safe, but it does entail some degree of invasiveness. Fortunately, vocalizations provided an alternative. At about 18 months of age, males begin giving their gender-specific form of the trumpeting unison call, in which they stand erect, drop their wings to flash bold black primary feathers, and emit loud, piercing notes. The female version lacks wing flashing and contains shorter notes. By age two, calls are separable with 100 percent accuracy.

Once gender was determined, Patuxent personnel began assembling suitable pairs — again, not so easy. Like humans, each crane is an individual with unique preferences for mate selection. Workers found that social status factored significantly into this equation — dominant high-ranking males tended to prefer high-ranking females, and vice versa. A high-ranking female would never submit to the attentions of a submissive male. Very low-ranking females often failed to form any pair bonds whatsoever. In addition, there were more obscure choice criteria involved; sometimes, two individuals just did not like each other. Occasionally there existed a crane that did not like anyone. One notorious bird at Patuxent, subsequently named Killer, murdered two females within several weeks of one another. Killer was improperly imprinted on humans, and perhaps this influenced his behavior toward other cranes. After these incidents, Killer was condemned to a life of bachelorhood, for good reason.

Aggressive behavior between cranes serves to establish and maintain social structure, an important component of flock stability. In the wild, however, birds can escape the wrath of an overly antagonistic individual. Not so in captivity — during Patuxent's early years, about one-third of adult sandhill crane mortality was due to aggression between birds in adjacent pens. The research center responded to this problem by designing a double-pen system where cranes were typically flanked by empty pens. These extra compartments were also useful for introducing potential mates or socializing juveniles with adult birds without direct contact.

All the courting, calling and rushing about functions primarily to get the pair in the mood. At Patuxent, this was also assisted by light-modification regimes that simulated conditions at Wood Buffalo. Despite these preparations, it was generally fruitless to just let nature take its course. Captive cranes are often tenotomized or pinioned — one wing is altered to impair flight — to prevent escape or injury within their pens. Regrettably, this hampers a male's ability to successfully mount and balance on the female's back during mating, and the wild flapping that ensues often infuriates the female into aggressive action. Consequently, most chicks produced in captivity are conceived through artificial insemination. Eggs are also routinely removed from nests because this induces females to lay more eggs, sometimes four or five times more than wild birds. Captive birds typically live much longer than wild birds in the predator-free and food-plentiful environment, so enhanced reproductive potential is truly the key benefit of propagation programs.

It is the hope of every captive breeding project that animals can ultimately be returned to the wild — *wild* being the operative word here — because instilling or maintaining wildness in another species requires insight, luck and considerable imagination. Of course, same-species parents are the best choice; however, a pair cannot possibly raise the half-dozen or more chicks that may hatch in a season. Other alternatives can engender improper sexual imprinting, where chicks have difficulty pairing with members of their own species once they mature. Cross-fostering is sometimes a less risky choice, particularly when young are raised by a closely related species. Whooping cranes at Patuxent have often been cross-fostered to sandhill cranes. Inappropriate sexual imprinting, has been avoided by removing young whoopers from their foster parents and placing them with juveniles of their own species before they pass through critical imprinting periods. Some sandhill crane parents proved better than others, but only the best earned the opportunity to raise a whooping crane. Years later, Ghostbird, an anomalous but handsome gray and white whooping-sandhill crane hybrid, would shine among the finest fostering birds as he both defended and nurtured his young adoptees with all the vigor of a true parent.

Other imprinting problems could be reduced by simulating a natural rearing environment using stuffed adult crane models that emit brooding calls. In addition, crane-head hand puppets can be manipulated to feed chicks either manually, by offering small pieces of food, or by mimicry that encourages young to search for food themselves by pecking and probing the ground. Older chicks can similarly be fed if puppet-wielding handlers are draped in a shapeless white costume. Although they look little like cranes, they look less like humans, and this is a secondary benefit. If a captive-raised bird is to be released in the wild, it must fear humans. This can be instilled quite simply by subjecting young cranes to disturbances, such as mock attacks, yelling and banging pots, all produced by uncostumed humans. Fear of nonhuman predators can be effected by using a trained dog to chase fledged birds; playing adult crane alarm calls at the same time increases the impact considerably.

These were by no means the only considerations when raising whooping cranes in captivity; every day would present research workers with new challenges. Even so, Dawn, the first chick produced by captive parents at Patuxent, hatched in the early morning on May 29, 1975. Just like Aransas Rusty, born 25 years earlier, she provided a big boost to staff morale. Unfortunately, little Dawn was born with a congenital foot deformity, causing it to twist outward; foot and leg abnormalities are common among captive-reared cranes and may be caused by genetic problems, among other things. Dawn refused to eat and died at just 15 days of age. The following year, two females (including Dawn's mother) produced five fertile eggs, from which one chick was successfully hand reared. It was a slow start, but within a decade, 197 eggs were produced by captive whooping cranes at Patuxent. One hundred and twenty-five eggs were kept in the captive propagation program; the remainder were taken off to greener pastures.

Meanwhile, the wild crane flock was still struggling through its annual cyclical dance and, although numbers increased through the 1970s, it was a bumpy road with no steady period of gain. Nonetheless, 57 Wood Buffalo–Aransas birds in autumn 1970 increased to 76 individuals by the close of the decade. During this time, there had been many changes on the wintering ground. Refuge workers began burning off patches of overgrown upland habitat, just as Mother Nature had done centuries ago, which increased quality crane foraging areas because it reproduced their preferred prairie-like habitat. Furthermore, management pushed to reduce the number of grazing cattle on the refuge. By 1970, there were more than 2,500 head destroying native vegetation and trampling fragile tidal zones. Fees per head were raised and numbers declined by two-thirds within 18 months.

Military activates on Matagorda Island had also decreased significantly, and the State of Texas was anxious to make use of land now lying fallow. Negotiations began, and by 1973, half of the barrier island was under the jurisdiction of Aransas National Wildlife Refuge. This decision's apparent stability was shattered a year later when the Air Force decided to close Matagorda facilities entirely, thus providing inroads for the Department of the Interior to take charge. A push-pull began between state and federal authorities that would last a decade and a half while refuge workers feared the worst: commercial and recreational development. In the end, these fears were not realized. Recreational development would not encroach too heavily on crane territories and, moreover, the Aransas refuge complex would increase in size to more than 115,000 acres (4,650 hectares), including the Blackjack Peninsula, Matagorda Island and some adjacent lands. Even the lands of rancher and oilman Toddie Lee Wynne eventually became part of the reserve, but not before Wynne, in collaboration with Houston-based Space Services Inc. of America, set out to prove that private enterprise had equal rights to conduct business in outer space. On September 9, 1982, they launched a rocket carrying a 1,000-pound (450 kg) mock payload from Matagorda Island. It traveled up 195 miles (314 km) before splashing down in the Gulf waters 10 minutes later.

So began a new epoch of crane research at Aransas. Audubon Society biologist David Blankinship led detailed studies of winter territorial behavior and feeding preferences. He specifically evaluated the effects of 30 years of shell dredging, oil and gas production and recreational and commercial use in and around the reserve on the quality of staple food organisms blue crabs, clams and mud shrimp — and more opportunistically-used species, such as snakes, eels, acorns and wolfberries. The 1970s were truly the beginning of the end of the "let them fend for themselves" era. It was a shame that Bob Allen was not there to see it.

But the decade was not without its setbacks. When 13 cranes failed to arrive in Aransas in 1972, Blankinship flew down the Gulf coast into Mexico, hoping that they were wintering elsewhere; he never found them. Migrating subadults were still taking the biggest hit; 16 Wood Buffalo pairs had produced 31 eggs that year, but only five juveniles made it to the wintering grounds. Maybe it was the weather during migration, and maybe it was not. Three of the five young disappeared over winter and the global population dropped to 49, where it held fast for two more years. The rollercoaster ride continued when optimal conditions in the north allowed 30 young to be added to the flock between 1975 and 1977; the numbers soared to 72 birds. It was then that research workers on both fronts realized that everyone would benefit from greater accountability.

As a result, the Canadian Wildlife Service began banding well-grown flightless young with colored bands to facilitate individual identification so that growth, seasonal movements, reproductive activities and longevity could ultimately be tracked. Each crane was fitted with plastic bands in various widths and color combinations distinctive to each bird — blue and white, red over blue

and so on — above the tarsometatarsus–tibiotarsus joint ("backwards knee") on one or both legs. Capture before fledging reduced undue trauma from the chase. Consequently, fatalities were few, numbering only two in 117 chicks banded between 1977 and 1987. Follow-up aerial surveys assured field-workers that the procedure was not disrupting family groups, and that banded young were not succumbing to greater predation. Banding was also yielding results on the wintering grounds. Aransas workers could easily track territory use, social interactions and winter survivability. Finally, the efforts of dedicated

Banding programs, led by Canadian Wildlife Service biologists at Wood Buffalo National Park, helped monitor whooping crane survivorship on both the breeding and wintering grounds.

Ron Sauey (left) and George Archibald (right) established the International Crane Foundation in 1973 with a promise of stewardship of the world's 15 crane species.

biologists at Wood Buffalo and Aransas were paying off. The wild flock continued to increase through the 1980s and never looked back. By the end of the millennium, this long-suffering company would measure 188 cranes strong.

Even in light of success, it is never a good idea to have all your eggs in one basket, or even two, when endangered species are concerned. During the early 1970s, there existed only two flocks of whooping cranes — one in the wild and one at Patuxent Wildlife Research Center. Home for a third group would soon be established in the rolling hills of Wisconsin, where the trumpeting calls of whooping cranes had been silenced almost a century before. Here, the International Crane Foundation was born of the hearts and minds of Dr. George Archibald and Dr. Ron Sauey.

George Archibald is the son of Nova Scotia farmers. He admits that his first avian passions were the chickens on his parents' New Glasgow farm. Archibald was saddened by the birds being so encumbered by life on land, so his premier research program, at age five or six, was attempts to make poultry fly. None were successful, of course. Furthermore, the chickens were so continually disturbed at their roost that they refused to lay eggs, and many ended up on the dining room table. From chickens, Archibald graduated to ducks and geese. He fussed continuously over "his birds" to the annoyance of his siblings and anyone else who attempted to cross the yard to milk the cow. One of his geese, obstinately nesting on the path, vociferously defended her turf without relenting. Archibald's interest in birds never waned, even in high school. When a local wildlife office initiated a ring-necked pheasant program, he bought an incubator and hatched birds in his bedroom. Archibald saw wild cranes for the first time in 1966, at age 20, while working in Alberta. He would later recall, "I saw the wilderness through new eyes, a place of life."

Archibald attended Dalhousie University in Halifax, destined for medical school until a chance meeting with ornithologist Dr. William Dilger, a professor of ethology (study of behavior) at prestigious ivy-league Cornell University, changed his destiny. After graduating from Dalhousie in 1968, Archibald headed south to Ithaca, New York, with less than a thousand dollars to his credit to pursue his doctoral degree. His Ph.D. dissertation would describe the evolutionary relationships between crane species based on their vocalizations. The board was set.

As a child, Ron Sauey was fascinated by all wild things, and Nodoroma, his country home in Baraboo, Wisconsin, was the ideal setting for these early outdoor experiences. He collected bugs and butterflies and raised pheasants in the yard. When he was seven years old, he proudly presented his mother with what turned out to be a not-so-dead possum. At 11, he made his place in the world as a field biologist by secretly gathering specimens for his older brother's school science project; rumor has it that he collected material for most of the class. It was shortly after this that his parents knew he would not be following his brothers into the family plastics business.

Sauey realized that he wanted to be an ornithologist before he could spell the word, and this ardor was first fostered by Baraboo High School biology teacher Gerald Scott. Scott was a true believer in learning from life, and on field trips they found eagles' nests, rescued a sick barred owl and observed wild sandhill cranes. That was a prophetic day, although Sauey would not know that for many years. After graduation from the University of Wisconsin, he was accepted into the doctoral program at the school where everyone who dreams of being an ornithologist wishes to go — Cornell University. He had planned to study pheasants with Professor Dilger until he met another of his students, one who was crazy for cranes. His interests quickly changed.

Archibald and Sauey decided, quite simply, over coffee one day that they wanted to save the worlds' cranes, and with this pledge they became ambassadors for these birds worldwide. To answer this calling, they established the International Crane Foundation (ICF) in 1973 on the Nodoroma property in Baraboo. Although their dream was simply to house captive breeding pairs of all 15 crane species, their goals were loftier. Their enterprise would provide a platform for research and education, institute programs for conservation of species and their habitats, maintain faltering crane populations through captive propagation and, perhaps most difficult of all, reintroduce captive-bred birds into the wild. They both longed to hear a whooping cry over the Wisconsin hills.

In those early days, cranes were kept in converted horse stalls that led to 15 large outdoor grass-covered yards that each included a spacious skylit house to shelter them in harsh weather. Flight netting would later be added to roof the high-fenced pens so that the birds did not need to be surgically rendered flightless. This would allow for the possibility of "more natural" breeding. The old pheasant coop became the chick-rearing facility. Two blind endangered white-naped cranes were the first birds to arrive from the Bronx Zoo; four ancient and bedraggled endangered red-crowned cranes were delivered shortly after — not an auspicious start, to say the least. Nonetheless, within two years the foundation had successfully hatched the first red-crowned cranes in the western hemisphere. Cranes continued to be added as the foundation's reputation soared. By 1976, Archibald and Sauey no longer had to beg people to send their cast-off birds; they had 12 species residing on the Baraboo property, five of which had bred successfully. One long-desired species not yet represented was about to arrive, and her name was Tex.

Tex was the 10-year-old human-imprinted female whooping crane, offspring of Rosie and Crip, that had been born at the San Antonio Zoo and tended by zoo director Fred Stark. She had been living at Patuxent for eight years but had been a nonstarter as a breeder. Consequently, she was transferred to the ICF in April of 1976, and a new potential mate named Tony (alias Georgette/George II, son of Josephine and Crip) arrived shortly after from the Audubon Park Zoo. But Tex did not like Tony; she preferred George Archibald. For all these years, her penchant for dark-haired men had been preventing her from responding to other cranes' attentions.

Archibald knew that captive populations could not afford to lose Tex's genes. Each bird represented a minute amount of genetic variability that was critical for the species' long-term maintenance both in the wild and in captivity. Of course, Tex could be artificially inseminated; however, she would not produce a fertile egg unless her hormones were jump-started by courtship. Archibald would have to win her heart. He moved a bed and desk into Tex's indoor pen, which was already subdivided by a wire barrier. He became her constant companion, standing guard while she ate or slept. In return, she defended him from "human" intruders by strutting about, pointing her formidable bill toward the ground and flashing her red head. Throughout summer and the following spring, Archibald danced with Tex. When she approached breeding condition, he helped her build a nest from grass and sticks. In mid-May, Tex laid her first egg. Sadly, it was infertile; Tony had proved a less than perfect donor of his genetic material.

The ICF searched for a replacement for Tony, while Archibald maintained his relationship with Tex. She would have no other, not even a stand-in mate

when he was out of town. Although she was unsure of her specific identity, at least Tex knew how to be monogamous like any true crane. It would take seven years of waiting, hoping and dancing before Tex would successfully fledge a chick. Gee Whiz was born June 1, 1982, and, although his early days were a struggle, he soon thrived under the indulgent care that he received. Days later, Archibald received an invitation to appear on the *Tonight Show* with Johnny Carson; he left for Los Angeles on June 22. Hours before the show was scheduled to tape, he got a call from Wisconsin. Tex's disemboweled body had been found in her pen, killed by a raccoon concealed in the discarded Christmas trees that had been used to block sight and sound in neighboring crane pens. The show went on and Archibald told his story to 22 million viewers. Messages of condolence poured into ICF offices the following day; however, few words can provide comfort in the passing of one's mate, even if she was of a different species. Yet Gee Whiz lived on as Tex's memorial, the hope for the future by virtue of the genetic lineages that he represented. He would father a generation of whooping cranes.

By the early 1980s, the ICF had outgrown its old digs and moved into a larger, 225-acre (90 hectare) site not far from Nodoroma. Over the decade that followed, they initiated extensive community-based habitat protection and

Whooper-girl Tex, imprinted on humans as a chick, takes an afternoon stroll with her chosen "mate," George Archibald.

restoration programs, raised global awareness regarding threatened species and safeguarded flyways for migrants. Furthermore, in 1985, they became the only facility in the world to house all 15 crane species. Two years later, Archibald and other ICF members were cut to the quick by the sudden, tragic death of Ron Sauey on January 7, 1987, following a cerebral brain hemorrhage on Christmas Day. He was 38 years old. Not since the days of Bob Allen had the world's cranes lost such an ardent supporter. But the cause went on; captive whooping cranes were destined for great things.

Discussions started as early as the late 1970s among George Archibald and other crane specialists favoring reintroduction of whooping cranes into central Florida. Difficulties associated with getting captive-bred cranes to migrate were irresolvable at this time, so it seemed a simpler solution to establish a nonmigratory flock. In the distant past, whoopers that bred in midwestern states wintered in Florida; museum records indicate that hunters dispatched 250 of these birds between 1870 and 1920. The last reported Florida bird was sighted on the Kissimmee Prairie in Osceloa County in 1936. Thus, this is where cranes would be released, on and adjacent to Three Lakes Wildlife Management Area, between Kissimmee and Yeehaw Junction. The resulting flock would be classified as "experimental nonessential," a designation under which the land used for the project would not be subject to restrictions, such as seizure, imposed by the Endangered Species Act. Local landowners were more willing to participate in the program if they did not feel that their rights were at risk. Biologist Stephen Nesbitt of the Florida Game and Freshwater Commission would be project leader.

The Florida site was selected because it was typified by open marsh-savanna country dotted by shallow lakes and palmetto (*Serenoa repens*) scrub — prime whooping crane habitat. There were relatively few humans but a great many cranes, including protected nonmigratory Florida sandhill cranes and wintering greater sandhill cranes, that could help new arrivals feel like cranes through good example. It was critical that any birds released to site be reared in visual isolation from humans by crane-costumed handlers to ensure that they did not suffer from any imprinting anomalies or lack the requisite fear of humans.

Breeding efforts to produce birds for the scheduled 1993 release began in earnest at Patuxent and the ICF in 1992. Also slated to join this crew were six chicks hatched from eggs transported from Wood Buffalo National Park that had also been raised in isolation. Chick-rearing facilities turned to veritable crane spas as humans draped in white robes sporting crane-head hand puppets led the young birds through their pre-release conditioning regimens, which included feeding, flying and swimming lessons, predator-avoidance exercises and early evening marsh walks.

Although conservation workers trapped and relocated bobcats (*Lynx rufus*) in central Florida, the predator remained the greatest cause of death among reintroduced whooping cranes.

The first cohort of 14 whooping cranes was released at Kissimmee Prairie in January 1993 at age eight to ten months. This was a time of increasing independence, when their wild parents, if they had had them, would be thinking about courting again. It was also after the first period of socialization, when foundations of pecking order are established, but before such animosity developed that fighting could produce injuries or fatalities. The cranes were retained in a 1-acre (0.4 hectare) conditioning pen to establish site loyalty before they were able to roam freely. It was critical that they remained faithful to the release site so that they could find mates readily when the time came, and avoid poachers that might be lurking beyond the protective boundaries.

Despite great attention paid to fostering wildness in the captive-raised cranes, almost half fell prey to bobcats within two months. Researchers determined that the young birds did not roost in open water surrounded by a functional "moat" that would impair mammalian predators. Wild cranes typically select sites with water about 16 inches (40 cm) deep, such as on a submerged sandbar. Even worse, the released birds tended to retreat into vegetation when threatened. Conservation workers trapped and removed 11 bobcats from a 1-mile (1.6 km) radius of the flock.

The young birds faced other perils, as well. When one bird appeared sickly, a veterinarian examined it for evidence of peritonitis; birds often injure their digestive system when they ingest bits of shiny metal that attract their attention. In this case, the vet found a treble fishhook lodged in the crane's throat. Some birds would later succumb to eastern equine encephalomyelitis, a viral disease similar to West Nile virus typically seen in horses. Although native sandhill cranes have natural immunity, whooping cranes in Florida do not. The 1993 cohort were inoculated for the disease; however, this practice was soon abandoned when workers reasoned that it would be impractical to immunize all future releases and their wild offspring. To be fully wild and self-sustaining, the birds would have to squeeze though this bottleneck.

Fourteen cranes were released into the wild during the winter of 1993; only five survived. The following year, only 5 of 19 released birds survived. Bobcat predation was the greatest cause of death. New protocols were instigated to encourage water roosting; mobile training pens were built to enclose suitable roost habitat and costumed handlers led the young birds into the water before darkness fell. Perseverance paid off and, by 1996, mortality rates had dropped from over 70 percent to about 30 percent. This success was also due, in part, to private landowners making additional habitat available to optimize these new release methods.

By early 1997, the first surviving birds were approaching sexual maturity. Refuge workers crossed their fingers as one pair was observed building nesting platforms along the edge of Hart Lake. However, this pair never had a chance; the male disappeared in June and was never seen again. A substitute male was captured and introduced to the lone female but they failed to bond with one another. When the male left to return to his original home at Three Lakes Wildlife Management Area, the female joined a flock of sandhill cranes. She was later found shot dead by hunters.

The first eggs of the Florida project were produced in 1999, but it was not until June 7, 2002, that a wild-born Florida whooping crane fledged — he was named Lucky, for all good reasons. Lucky survived drought, eagle attacks and boys on bicycles; the following year, his parents raised Lucky II. Dedicated workers at the International Crane Foundation and Patuxent Wildlife Research Center had shown the world that they could teach captive-bred whooping cranes how to be cranes and successfully release them into their natural habitat. If only they could teach them to migrate, then the birds would be truly wild.

Chapter 4

Recovery of the Whooping Crane:
Migration

Ornithologists define avian migration as *regular movement of birds between areas inhabited at different times of the year.* Yet how poorly this banal statement articulates the vast seasonal associations of winged travelers that have occurred through the millennia with regularity almost as dependable as the oceans' tides. In their need to explain the cyclical appearance and disappearance of local populations, humans often endowed migratory birds with mystical qualities. Aristotle believed that some species were capable of transmutation: Europe's summer redstarts changed into winter robins, and garden warblers became blackcaps. He professed to have seen these species in the very midst of re-creation, when their feathers claimed characteristics of both species. We now know that Aristotle was merely witnessing birds in prebasic molt, which gradually transforms showy breeding plumage into more somber, less conspicuous winter colors. Other early naturalists explained the seasonal absence of birds through miraculous hibernation. Twelfth-century Welsh monk Giraldus Cambrensis suspected that barnacle geese (*Branta leucopsis*) — seen only during winter in Britain — spent their summer more profitably as gooseneck barnacles (*Pollicipes polymerus*), small, sessile crustaceans of similar color and shape. Invoking such supernatural mechanisms to explain migration seems ludicrous to us now. However, we must remember that it was centuries later — after humans had invented practical means for global travel — that explorers and naturalists began to return to their homelands bearing bird skins from exotic locations that told the story of where "our" birds were when they were not with us.

Birds are by no means the only migrant animals — some fishes, butterflies, deer, whales, bats and others also undergo regular seasonal movements — however, migration is pervasive among the world's almost 10,000 bird species primarily because they can fly. Evolution has certainly favored the origin of long-distance travel in the absence of formidable land-bound geographic barriers. Each autumn approximately 10 billion birds representing almost a thousand species leave their northern summer homes to winter at more southerly latitudes. Migration is not without cost, for indeed, more than half of the birds that depart their northern breeding grounds fail to return the following spring.

Migrating Eurasian cranes congregate on Hula Valley wetlands in northern Israel to rest, feed and socialize before continuing their southbound journey to wintering grounds in Africa.

Some die on the wintering grounds but most meet their demise from starvation, predation, exhaustion, inclement weather, collisions with human-made structures and hunting somewhere en route. Yet migration would not be so widespread among birds unless the benefits of seasonal movement broadly outstripped the costs.

Most migrant birds move from temperate breeding sites to warmer wintering habitats, often at tropical or subtropical latitudes. Even in the Southern Hemisphere, some species — known as austral migrants — fly north toward the equator to find more favorable environs in winter. However, they are rare, relative to Northern Hemisphere migrants, due primarily to the paucity of temperate land masses in southern oceans. In other world regions, such as sub-Saharan Africa, where seasonal climates are best delineated by precipitation, some birds move regularly with the rains to ensure adequate resource availability.

Migration is best seen as a successful strategy to exploit seasonally favorable resources, not merely one designed to avoid intolerable conditions. Temperate latitudes offer food supplies that fluctuate seasonally, being typically low in winter and high in summer; abundant food during the breeding season allows parents to provision their chicks quickly and with less effort. Although the tropics may have more predictable food sources year round, birds are long-lived and the moderate climate favors high seasonal survivability. Thus, tropical birds breed at a much higher density and, consequently, competition for food and nesting territories is fierce. Many birds, including cranes, will not breed unless an appropriately appointed territory can be maintained. Securing a tropical breeding territory can be compared to finding an apartment in a building where no one dies and no one ever moves out.

During the nonbreeding season, however, many tropical birds relax their territorial hold, allowing northern migrants to bask in the moderate weather and accessible food that characterize their southern winter homes. But as a distant spring approaches, these seemingly perfect quarters are once again too small. And although the long trek north presents migrants with many perils, the promise of wide-open spaces and vast blooms of seasonally abundant food is relentlessly alluring. This is why most birds migrate. In fact, many ornithologists believe that the northern birds that briefly grace our forests, backyards and fields in summer are not "our" birds at all, but, rather, migratory descendants of sedentary southern species that have been selected through evolutionary time to seek out more profitable, yet remote, breeding habitats. Ornithologist Dr. George Cox coined this "southern ancestral home" hypothesis to explain the disparate summer and winter distributions of many ex-tropical birds, such as wood warblers, hummingbirds and flycatchers. But regardless of their specific origins, it is important to remember that migratory birds are not "visitors" in any sense. They are ecologically important components of both their summer and winter habitats — and all places that they gather in between — that have evolved intimately and in tandem with all the biotic elements that comprise these habitats. Their mark is indelibly stamped across their world.

The lengths of avian migratory routes vary considerably, as evolution has shaped the annual cycle of each species to meet its particular ecological needs. Some birds, such as the Arctic tern (*Sterna paradisaea*), seem restricted only by the limited size of our planet — their annual route takes them from one pole to the other, a round trip of some 25,000 miles (40,000 km). Even diminutive shorebirds such as red knots (*Calidris canutus*) regularly cover more than 8,000 miles (13,000 km) one way from Baffin Island, in northern Canada, to Tierra del Fuego, at the southernmost tip of South America. The migrants' routes often take them over all manner of habitats, be they hospitable or not: deserts, oceans, forests and grasslands. A few species fly nonstop, but most make periodic stops. Songbirds typically migrate under cover of darkness, a strategy that obscures them from hungry diurnal predators, such as hawks, and allows them to feed during daylight. Stopovers typically occur where food is plentiful, because migratory birds often need sustenance more than they do rest. Birds may linger several days in more profitable locations until favorable weather and the urge to travel sends them on their way again. Generalistic species may find many such locations; however, birds with more discerning tastes resort to only a few choice stops on their entire route. Their expectation is that these traditional-use sites, knowledge of which is passed down through generations, will be available unchanged each year. When they are not, the result can be disastrous.

Unlike songbirds, whooping cranes migrate during the day; they have little fear of predators in flight and their large, broad wings are best fashioned for soaring during the heat of the day. Their semiannual trek covers about 2,400 miles (3,860 km) between the Northwest Territories and the Texas Gulf Coast along well-traveled migratory routes; radio-tracking studies have demonstrated that a family's flight path may deviate only about 30 miles (50 km) on subsequent journeys north and south. Whooping cranes, particularly males, also exhibit strong natal philopatry, an innate behavior that favors their return to nest sites near where they were raised. The benefits of revisiting habitat of proven breeding success are obvious; also, repeated use of traditional territory ensures profitable knowledge of the site's resources — it is easier to find the pantry in one's own home than in a neighbor's abode.

Predictability of movement, however, has made cranes more vulnerable historically, particularly during stopovers where they encounter humans and human-modified environments. Moreover, cranes demonstrate less opportunism en route than many other species. The Platte River mudflats in southern Nebraska are traditional stopover sites for both sandhill and whooping cranes because abundant food and safe roosts offered along the river are crucial to success on their long journey. Unfortunately, prime river habitat is being threatened by development, particularly upstream damming that modifies flood cycles and favors invasion of woody vegetation downstream. Some ornithologists believe that this incremental reduction in critical stopover habitat may be indirectly linked to limited population growth in whooping cranes.

This is by no means the only human-induced risk associated with crane migration. Critical habitat has also fallen to land-leveling equipment and center-pivot irrigation systems, and government subsidies that have exacerbated the loss through funding so-called land-improvement projects. Wetlands have also been contaminated by oil and gas drilling, and by surface mining that causes an increase in toxic trace elements, including mercury, lead, arsenic, beryllium and cadmium, in water and surrounding soil. In addition, agricultural runoff of fertilizers and pesticides causes both mortality and increased reproductive failure. As prime stopover habitat is reduced, incidence of infectious diseases such as avian cholera increases among densely packed roosting birds. Migrating cranes, particularly juvenile whooping cranes, are also frequent victims of lethal collisions with towers and utility lines that crisscross their travel corridor or surround their traditional roosts. But, perhaps, hunters have had the greatest opportunity to exploit predicable habits inherent to crane migration — they know precisely where to find the birds at given times of the year. Migration has always been dangerous, but it has become infinitely more so since humans walked the earth. This perhaps explains why it is the migratory species among all cranes that are in the gravest danger of extinction.

Despite natural and human-caused risks, migration is a worthwhile strategy, and evolution has thus shaped birds to maximize their chance of completion. But successful migration requires two different interconnected skills: finding the way and being physically able to get there. One must not underestimate the importance of either — birds' arduous journeys are unlike anything humans can do unaided by modern machinery. Nomadic peoples of African deserts or the Arctic may move a few hundred miles each year; the diminutive common swift (*Apus apus*) flies over 600 miles (1,000 km) per day. Even the tiny ruby-throated hummingbird (*Archilochus colubris*) makes a lonely 500-mile (800 km) nonstop flight across the Gulf of Mexico in about 20 hours, arriving in northern South America on the very last reserves of its energy.

In all birds, the semiannual urge to migrate is mediated internally by hormones that are influenced externally by factors such as day length. Physiologically, these changes bring about *zugunruhe*, or migratory restlessness, which simulates birds to feed excessively, which increases fat reserves required to fuel their journey. Body fat is the primary source of energy for most migratory birds, except those few species such as swifts, swallows and some seabirds that can feed on the wing. Prior to migration, birds feed voraciously on energy-rich foods, building up substantial fat deposits subcutaneously (under the skin) and within their muscles and body cavity. Fat reserve requirements vary greatly among species, depending on their size and stopover opportunities. Nonmigratory birds typically carry 3 to 5 percent body fat, and short-range migrants that stop frequently have 13 to 25 percent. However, small long-distance migrants that fly long stretches nonstop, such as plovers and other shorebirds, must double their body weight in several weeks of ravenous feeding prior to departure. When metabolized, these fat reserves are an excellent fuel source, yielding twice as much energy and water per gram than proteins or carbohydrates. If a blackpoll warbler (*Dendroica striata*) were burning gasoline instead of body fat, it could boast a fuel consumption rating in excess of 720,000 miles per gallon (303,000 km/l).

Large birds, however, burn fuel much more quickly due to their greater weight and bulkier flight muscles required to keep them aloft. Flapping flight requires considerable energy, and can rapidly deplete valuable energy reserves. Typically, large birds are unable to store sufficient fat to power continuous flapping flight and must rely on regular stopovers to refuel. Furthermore, migratory flight in large birds such as cranes is often a blend of powered flapping flight, thermal soaring and gliding.

Soaring and gliding are used extensively as energy-saving measures, if appropriate circumstances prevail. Cranes have large, high-lift slotted wings that allow them to ride effortlessly on thermal air currents, which are rotating columns of rising warm air that are formed over the earth's heated surface. Thermals typically set up in predictable geographic locations, often above hills and escarpments, which is why most large, broad-winged migrants adhere to traditional migration corridors. These migrants wait until the air is sufficiently heated before they depart, typically in the morning or early afternoon. At first, they flap-fly until they find an active thermal current. Once in a thermal, the birds can soar on set wings provided that they continue to turn within the rotating column as they gain altitude. Cranes frequently climb 6,500 feet (2,000 m) or more in a thermal before they leave it, turning toward their ultimate destination, and gliding unidirectionally as they gradually lose altitude. Before reaching the ground, they find another thermal and repeat the circling and gliding process. Many species travel over land throughout the day in this way. Soaring on thermals saves substantial energy, although it may take more time to cover distance.

If weather conditions, geological formations or bodies of water prevent the creation of thermals, cranes migrate in a V-formation to conserve energy. During flapping flight, air flowing over and under a bird's wing — the flow that itself creates lift — also contributes to turbulence near the tips of the wing's trailing edge. These whirling eddies of disturbed air, called wing-tip vortices, provide a bird following just off a leading bird's wings with a little added lift, thus saving some energy in powered flight. When flying in V-formation, all birds except the leader benefit, which is why flock leaders change frequently when migrating.

Cranes also save energy when migrating by choosing flying altitudes with favorable wind conditions. Typically, winds at higher altitudes are stronger than winds closer to the earth's surface; therefore, the benefit of a tailwind (a wind blowing in the direction that the bird is heading) will be increased if the bird flies

Flying in V-formation, particularly during migration, saves considerable energy because following cranes receive some added lift off the leading birds' wingtips.

higher. Cranes rarely migrate into headwinds; however, if they do, they typically fly at much lower altitudes. Sandhill cranes have been reported at altitudes of almost 12,000 feet (3,600 m), but typically fly no higher than 5,200 feet (1,600 m). Eurasian cranes, on the other hand, have been observed at altitudes of 33,000 feet (10,000 m) when migrating over the Himalayan Mountains.

Flight speeds and daily migrating distances have been poorly studied in cranes; many species probably migrate at average overland speeds of 35 to 50 miles per hour (60 to 80 kph). Studies of migrating whooping cranes indicate that they generally travel 110 to 200 miles (180 to 320 km) per day in about three hours' flying time. This daily distance and flight time may be typical, although favorable wind and weather conditions can increase this significantly. For example, sandhill cranes have reportedly flown 360 miles (575 km) without stopping to rest, and individual flights of 500 miles (800 km) per day have occurred. These distances, however, may not reflect those of family groups migrating in autumn, which travel more slowly with young in tow. Regardless, each trip between wintering and summering homes undoubtedly takes many days to complete.

Timing of migration is critically important, particularly in spring. Early arrival on the breeding grounds better guarantees optimal habitat choices and provides a longer period of time for raising young before the next autumn migration. However, birds that arrive too early may find little food available to keep them alive. We all fear the effect of late snows and cold weather on the first robins of spring. In autumn, departure from the breeding grounds is typically more leisurely, provided that food resources do not decline drastically. Pairs with late clutches usually leave much later than others to allow adequate time for their young to mature before their journey begins.

Departure is always mediated by weather; most birds migrate only when weather conditions are favorable. Autumn migration is typically benefited by low-pressure systems traveling across the North American continent from the west in a general easterly direction. These systems rotate in a counterclockwise direction, and brisk northerlies produced as they pass through provide strong tailwinds. In spring, birds wait until frontal systems create strong winds from the south to ease their passage. Soaring birds must keep an additional weather eye for thermals, which do not form readily in cold, wet weather. Exactly how migratory birds forecast weather remains a mystery. Certainly they are able to sense changes in barometric pressure; birds feed more intensely at backyard feeders as the barometer falls. Studies have also suggested that they may be able to detect naturally generated infrasound (ultralow-frequency sounds) that emanate from turbulent air masses.

On departure day, whooping cranes demonstrate increased alertness. They feed intensively and preen in anticipation. Adult males repeatedly tilt their heads to look up at the sky, perhaps to assess weather conditions. This curious behavior is observed regularly — every half-hour or so — pending migration, but is rarely observed prior to local flights. Just before takeoff, the small family flock clusters together and faces into the wind with the adult male at the head. He leans forward, advances a few steps and becomes airborne just an instant before other members of his clan follow suit. Although whooping cranes migrate primarily in family groups, they are not above encouraging others to depart. As small flocks take to the air, they may linger a while, circling perhaps 500 feet (150 m) above the ground over other cranes, calling to them with their bugling voices. Sometimes these actions will cause numbers of trumpeting birds to take flight simultaneously, circling upward and northward together, until finally they separate into smaller bands that will find their own way to their remote breeding grounds.

If making the journey seems unfathomable to us, then understanding birds' ability to find these distant lands, with no want for a road map, seems even more so. Finding the way is also a twofold challenge: migrants must possess the knowledge of where they are and where they are going, and they must have the means to navigate along the route between those two points. Avian navigation is as varied as the many modes of migration. Certainly birds that travel during daylight hours make good use of visual landmarks inherent to the landscape by following rivers, shorelines and mountain ranges. Birdwatchers frequently take advantage of this by gathering along prominent landforms, such as peninsulas at Point Pelee, Ontario, and Cape May, New Jersey, where migrants are funneled through restricted travel corridors. Diurnal migrants also navigate using a solar compass that is time-compensated to account for the sun's changing position from sunrise to sunset. Adélie penguins (*Pygoscelis adeliae*), which walk or swim along their 380-mile (600 km) migratory route north from the Antarctic coast, correct their orientation by 15 degrees per hour relative to the sun's position, indicating that they understand the apparent counterclockwise movement of the sun in the Southern Hemisphere.

Birds that migrate at night use the stars to guide them. Like diurnal migrants, night fliers use a time-compensated compass to assess their orientation relative to the stars' changing positions. Ornithologists once believed that birds set their course using only a single bright object, such as the North Star in the Northern Hemisphere sky. However, orientation experiments using indigo buntings (*Passerina cyanea*) have demonstrated that they know the pattern of major constellations within 35 degrees of the North Star, including the Big Dipper, Little Dipper, Draco, Cepheus and Cassiopeia. If clouds obscure part of the night sky, then they merely select another constellation to guide them. Likewise, birds of the Southern Hemisphere orient on constellations surrounding the Southern Cross. We can only imagine the wealth of astronomical knowledge inherent in the minds of Arctic terns, whose circumpolar migrations require that they appreciate constellations in both hemispheres.

But birds also travel in cloudy weather, when solar and stellar compasses are of little utility. At these times, they rely on less obvious means to guide them home: olfactory cues such as the smell of the sea, infrasound generated by wind traveling over land, and the earth's magnetic field. It is apparent that our reliance on instrument navigation may pale in comparison to the innate sense of direction in migratory birds. One afternoon in central Illinois, a gray-cheeked thrush (*Catharus minimus*) was captured in a mist-net. Researchers, anxious to follow its migratory route, attached a tiny radio transmitter to it. At dusk the thrush departed on the next leg of its journey with the ornithologists pursuing in a small plane. A severe thunderstorm and dwindling fuel supply forced the plane down during the night. The thrush, however, landed safely in Wisconsin at dawn the next day. It had traveled 400 miles (650 km) nonstop through the storm holding a firm compass bearing.

The ability to navigate appears to be partly innate and partly early learned behavior, and avian species vary substantially in this regard. It appears that all birds have some inborn map of the world — in order to get somewhere, they must know where they started. Most birds can find their way back to the breeding grounds where they were hatched after their first winter away. Migratory birds experimentally displaced to the west know to fly east, and south to north and so on. Homing pigeons fitted with tiny frosted contact lenses, which made them severely nearsighted, flew over

100 miles (170 km) directly back to their coops and landed like miniature helicopters on the roofs. They knew where to go without ever having traveled the route before.

Ornithologists are unsure exactly how a bird's internal map is generated. It is, perhaps, somehow connected to solar and stellar compasses that provide information about latitudinal and longitudinal position. Marine navigators have historically used a sextant — an instrument that measures the angle of elevation of a celestial object above the horizon — to plot a position line on a nautical chart. Birds may do much the same. Planetarium experiments have shown that chicks in the nest learn the stars during their first months of life as they gaze sleepily up at the night sky. Perhaps they set their solar compass during the day. This knowledge may be calibrated to directional information derived from the earth's magnetic field during their first migratory flights. If a bird has a sense of what direction to travel and how long it requires being on the wing, even the most inexperienced traveler should reach part of its species' distribution. Of course, navigational abilities improve with age. Most accidentals — birds reported outside of their typical distribution — are young of the year.

Some doubters merely dismiss the magic of avian migration, claiming that birds learn everything from their parents or other birds they accompany on their first flights. However, these cynics should consider the navigational

Almost half a million migrating sandhill cranes stage on a 75-mile (120 km) stretch of southern Nebraska's North Platte River, just as they have done every spring for the past 10,000 years.

abilities of New Zealand's shining bronze-cuckoo (*Chrysococcyx lucidus*). Like 50 other cuckoo species, the shining bronze-cuckoo is a brood parasite that lays its eggs in other species' nests, including the grey warbler (*Gerygone igata*). Adult cuckoos depart New Zealand soon after they abandon their eggs, leaving the foster parents to rear their young. They migrate about 1,200 miles (1,900 km) westward to Australia and then turn northward over the western Pacific to make landfall on the Solomon and Bismarck islands an additional 1,000 miles (1600 km) distant. This migratory feat is matched about one month later when the young cuckoos — having been raised by a nonmigratory species — depart the breeding grounds, fly the same route and join their parents on a tiny speck of land amid an otherwise featureless expanse of ocean. Yet they find it, having never been there before and having no one to show them the way.

Certainly the shining bronze-cuckoo demonstrates the highest degree of innate navigational ability; many other bird species have the opportunity to rely on their parents' migration experience. It has been known for some time that young whooping cranes accompany their parents on their premier migratory flight, thus acquiring some knowledge firsthand. However, traveling with parents has benefits other than just learning the way, such as increased vigilance for predators and improved stopover selection, that could easily direct the evolution of this mode of migration. What remained to be garnered about crane migration was whether or not learning the route was the primary function of traveling with a family group. Was a substantial part of navigational ability innate in cranes? This question was critical to the whooping crane recovery program if a second migratory population was to be established. And while a simple "yes" would have provided a comparatively effortless solution, it nonetheless would have deprived humanity of a great adventure.

Displacement experiments are commonly used to determine if migratory behavior is based on learned or innate intelligence. If a young bird is moved some distance from its natal territory and homes without prior experience, we can assume that a significant portion of the knowledge is inborn. If the bird does not initiate migration, or shows little clear directional movement, it is likely that this species requires a guide-bird — a parent, sibling, or other flock member — to provide a migratory template for these early learning experiences. This is, of course, an oversimplification, but the primary question remained regarding reintroduction of migratory whooping cranes. Would they know where to go?

Several experiments with sandhill cranes in Florida during the 1980s provided key information about crane migration. Initially, this work was designed to determine if reintroduced nonmigratory whooping cranes would stay put; straying too far from protected lands would have placed them in certain jeopardy. Between 1981 and 1984, biologist Stephen Nesbitt placed 19 eggs of migratory greater sandhill cranes into 12 nests of nonmigratory Florida sandhill cranes; 6 eggs had come from wild birds in Wisconsin and the others from captive birds at Patuxent Wildlife Research Center. Only two chicks survived; however, neither of these birds showed any inclination to migrate, even after cavorting with wild compatriots that passed through the reserve in winter. Other introduced birds followed suit. In 1986, Nesbitt released 15 parent-raised greater sandhills from Patuxent into northern Florida. Although the experimental birds drifted about 70 miles (112 km) south with the wild flock when they arrived to winter, they failed to travel north with them in spring. The birds did eventually move back to within 30 miles of their release site, perhaps showing some degree of natal philopatry.

This small glimmer of migratory knowledge, if indeed that was what this local movement indicated, would be critical to whooping crane reintroduction. Protocols to establish a second migratory flock would be based heavily on natal philopatry — if researchers could somehow compel the birds to travel the route once, they might come back to where they were introduced on their own volition the following spring. Once migratory knowledge was established in a population, then all future generations would have someone to show them the way.

However, reintroducing a viable migratory flock is not as straightforward as just guiding birds along a flight path. It is equally important that they be able to fend for themselves when human intervention retreats. Thus, ornithologists recognized three additional challenges to any such reintroduction program: preserving a sufficiently high survival rate among introduced birds; inducing fear of humans; and ensuring that this acquired knowledge is passed on to offspring through typical reproduction. Basically, researchers had to make wild cranes from potentially captive stock. In addition, the Florida sandhill crane experiments would demonstrate all too clearly that release methods could substantially affect wildness and survivability.

In the mid-1970s there were just over 50 birds in the Wood Buffalo–Aransas population, and thoughts of establishing a second migratory whooping crane flock were finally being put into words. But how could it be done? One plan proposed relocating wild adults to new summering grounds and placing them under flight constraint and supervision. Once they bred, their young would return to their natal area and, thus, establish a migratory population at this new location. A second initiative recommended using wing-clipped, captive-bred subadults placed in enclosures to attract wild birds as mates. After successful pairing and reproduction, the wild bird would teach migration to both mate and offspring. Finally, cross-fostering was proposed. This strategy entails removing young from their biological parents at birth — or still in the egg, as is the case with birds — and giving them to surrogate parents of a different, but similar, species to be raised. Cross-fostering is used regularly in commercial livestock facilities to increase production and has been attempted in other endangered species recovery programs, including the peregrine falcon (*Falco peregrinus*).

The notion of using sandhill cranes to cross-foster endangered whooping cranes was first formally introduced by Fred Bard at the Whooping Crane Conservation Association meeting in 1963. As director of the Royal Saskatchewan Museum, Bard had long been interested in whooping crane recovery. After all, Saskatchewan was on the migratory route, was home to the last recorded nests south of Wood Buffalo National Park, and had a morbid history of human-caused devastation to the species. Bard had worked actively in the 1940s with Robert Allen and John Baker of the National Audubon Society to publicize the crisis. Cross-fostering offered considerable potential, and the logical surrogate wild species for whooping cranes were the abundant sandhill cranes.

Both species maintain lifelong, monogamous pair bonds and typically use the same breeding territories every year. Their eggs are similar in size, shape and color and require about 30 days for incubation. Furthermore, chicks of both species are alike in appearance; they stay with their parents most of their first year and, best of all, travel with them on their first migration. Perhaps sandhills could teach naïve whoopers how to be cranes. Critics, however, were quick to note some significant differences between the species. Whooping cranes are more carnivorous, more aquatic and more solitary than sandhills. In addition, each species has distinctively different visual and vocal displays. Optimists were convinced that these differences would work to their advantage by preventing

Bosque del Apache National Wildlife Refuge is winter home to more than 35,000 snow geese and, for a time, a handful of cross-fostered whooping cranes.

potential hybrid matings once the two species were sassociating closely with one another.

The 15-year undertaking — a cooperative venture of the Canadian Wildlife Service and US Fish and Wildlife Service — would officially begin in 1975. Considerable thought was given to selecting a location for the project. The reintroduction site would need to be within historical whooping crane range, be geographically isolated from the Wood Buffalo–Aransas flock and provide extensive suitable habitat to serve present and future populations. Grays Lake National Wildlife Refuge in southeastern Idaho was ultimately selected.

Grays Lake NWR was established as a waterfowl reserve in 1965, and its 22,000-acre (9,000 hectare) marsh is an important nesting ground for greater sandhill cranes (*Grus canadensis tabida*). Males of this statuesque sandhill subspecies are almost as tall as whooper females. The primary distribution of greater sandhills is mid-continent; however, they also occur in isolated Rocky Mountain river valleys and meadows above 5,000 feet (1,500 m), from southern Oregon and northeastern California east to northwestern Colorado. Grays Lake cranes migrate about 850 miles (1,350 km) to winter at Bosque del Apache National Wildlife Refuge, south of Albuquerque, New Mexico. The reserve offers waterfowl and wading birds almost 25,000 acres (10,000 hectares) of managed wetlands and corn and alfalfa croplands, and supports wintering crane populations of 15,000 to 20,000 birds. Between Grays Lake and Bosque del Apache, migrating sandhill cranes regularly use stopover sites located in other federally protected areas, including Ouray NWR on the Green River in Utah and Monte Vista NWR in Colorado's San Luis Valley.

The lead researcher at Grays Lake NWR was Dr. Roderick C. Drewien. Drewien had been studying greater sandhill cranes for decades, and had based his doctoral dissertation on their ecology. The Idaho cranes showed considerable promise as foster birds. First and foremost, they were excellent parents. The 250 crane pairs at Grays Lake had unusually high nesting success, fledging 80 to 90 percent of their young. Moreover, Drewien had banded several hundred birds over the years, and knew which pairs were the most stable and reliable breeders. Also, many aspects of the region were similar to whooping crane nesting habitat despite differences in elevation — about 6,500 feet (2,000 m) at Grays Lake compared to 1,000 feet (300 m) at Wood Buffalo. The wetlands were also relatively pollution-free.

Perhaps the greatest sticky wicket would be hunting. Drewien and his team had tracked the sandhills' migratory path to assess potential risks en route for reintroduced whooping cranes, and had concluded that there would be little hunting pressure during migration. Wintering grounds in Bosque del Apache, however, were a mecca for waterfowl hunters, and whooping cranes look sufficiently like any one of 35,000 snow geese that visit the refuge during the hunting season. Moreover, the Central Flyway Waterfowl Council — whose members represent state and provincial interests regarding migratory game bird management — was adamant that no whooping crane reintroduction project curtail the hunting season. Nevertheless, Grays Lake's sandhill cranes would become foster parents for whooping cranes, and if luck smiled upon them, they would successfully incubate the eggs, raise the foster chicks and show them how to migrate to New Mexico. There was a lot at stake that first year: all 14 eggs had been taken from the wild flock at Wood Buffalo National Park, which numbered only 49 birds.

On May 31, 1975, CWS biologist Ernie Kuyt flew in to Grays Lake with his precious cargo of well-incubated eggs. Studies had indicated that one of two eggs could be removed from crane nests without any reduction in seasonal reproductive success; however, some biologists pointedly challenged the concept of the "redundant egg." After all, twins were occasionally observed at Aransas. Nevertheless, this spring these eggs faced a different destiny. Within six hours of arrival in Idaho, all 14 whooper eggs had been switched with an egg in resident sandhill nests. The second sandhill egg in some of these nests would be removed later; doting parents could count to two — sometimes — and Drewien did not want the nests to appear too "altered" because it could increase the likelihood of abandonment. An aerial survey the next day found 13 of 14 eggs being incubated diligently. By early June, nine eggs had hatched; three others were infertile and two had been taken by predators. The newborns appeared well cared for by their foster parents, who responded industriously to their begging calls. Moreover, the chicks were attending to their parents' alarm and brood calls appropriately by hiding and following on command. The families foraged together in upland grainfields, and all looked well.

By midsummer, however, things began to go amiss. Despite the apparent attention of foster parents, whooping crane chicks were simply vanishing. Coyote predation was certainly a problem, as was the snowstorm of late June. Grazing cattle were also taking their toll. Chicks frequently panicked as cattle approached, often separating them from their parents. Fleeing whooper chicks, bigger and gawkier than young sandhill cranes, could easily get snagged, sometimes fatally, in the miles of six-stranded barbed-wire fence crisscrossing the crane territories. Parents could fly over such obstacles; youngsters could not. The stumbling blocks faced by cross-fostered chicks at Grays Lake were exemplified by a bird that researchers appropriately christened Miracles. This resolute young crane had faced the barrage of perils in Idaho — inclement weather, barbed-wire fences, utility lines, cattle, coyotes, dogs, ranchers — and, unlike most other members of his cohort, Miracles had survived them.

When the sandhill flock departed on the journey south in mid-October, only four whooping cranes were left to go with them. Two of four foster families arrived safely at Bosque del Apache NWR after a lengthy stopover at Monte Vista NWR in Colorado. One family wintered on dairyland near Los Lunas; Miracles ended up at Bernardo State Wildlife Management Area, about 40 miles (65 km) north of Bosque del Apache. Once again, this bird defied the odds. Illegal crane hunting was rampant at Bernardo; when apprehended, hunters would claim they were shooting at "large gray ducks." Nonetheless, Miracles dodged illicit gunfire as he foraged in fields, and flew the gauntlet across the frontlines of legal hunters each day as he returned from roost. Eventually the corn ran out, and Miracles and his parents moved north.

Chick–parent behavior had been typical through early winter, although foster chicks were frequently harassed by other adult sandhill cranes. Also, families with foster chicks were never really accepted into sandhill society and tended to avoid confrontation by loitering on the periphery of winter-feeding flocks. Other behavioral problems began on the return trip. Both whooping and sandhill crane juveniles typically accompany their parents back to the breeding grounds before they strike out on their own. However, the cross-fostered cranes began to detach themselves from their familial unit in late February. Only one chick reportedly stayed with his parents during migration, and none of the four juveniles that had survived the winter returned to Grays Lake with them in spring. The young whoopers showed up in Utah, Montana and elsewhere in Idaho, and remained ostracized wherever they traveled. Only the most dominant individuals, such as Miracles, were able to feed among sandhills, but they were compelled to threaten, peck and chase the other birds to remain among them. More submissive young whoopers were never able to assimilate into juvenile sandhill flocks and chose to retreat from their space entirely. Two male fostered cranes joined forces briefly during the fall of 1976, but even this association was tenuous. They parted ways by mid-November. Something was dreadfully wrong, but still the experiment continued. By the spring of 1978, 61 whooping crane eggs had been fostered to sandhill cranes at Grays Lake. Only six birds had survived their first year of life. Eighteen more eggs would be placed under sandhill parents in 1978, but only three would fledge. Explanations remained elusive; most of the dead had just disappeared.

By 1978, the first cohort of cross-fostered whoopers was old enough to begin laying claim to summer territory in Idaho. The three survivors had finally established some regularity in their seasonal movements between Grays Lake and Bosque del Apache. However, this year only Miracles, most dominant of the three, was able to defend some 60 to 80 acres (25 to 30 hectares) where he had fed the previous summer. The other males, Pancho and Corny, would have some success in 1979 and 1980. These birds kept to themselves throughout the summer and winter, and showed little evidence of association with any other whoopers, including younger females.

Even if the males had been so inclined, mates were not overly abundant at Grays Lake. As a rule, single male whooping cranes are more philopatric than females. This is evolution's mechanism to mix genetic material — males home and females disperse. In a healthy population with sufficient individuals throughout their distribution, young roaming females would, thus, be introduced to territorial males that are not likely to be closely related to them. This would not work in Idaho because wandering females would rarely encounter another whooping crane. An additional problem was survivorship. Most of the vanished birds were females; their generally more subordinate demeanor caused them to be shunned by flock members, putting them at greater risk of starvation, predation, hunters and other calamities.

But the 1975 cohort was now of breeding age, and Drewien and his team were anxious to see if the cross-fostered birds would reproduce; this was critical to the project's overall success. Due to the shortage of females at Grays Lake, biologists from Patuxent and the Idaho Cooperative Wildlife Research Unit devised a plan to introduce a female in between territorial males to see if something would click. Unfortunately, only Miracles returned from Bosque del Apache in spring 1981. Pancho had vanished on the wintering grounds and Corny had disappeared during migration. Hunters or collisions with power lines were likely causes of their deaths. By default, Miracles became Romeo to a captive-bred Juliet named Too Nice.

Too Nice had been hatched at Patuxent in 1978 from an egg collected at Wood Buffalo. She was three years old when she was slated to meet Miracles, not quite breeding age but certainly old enough to consider bonding with an eligible male. Too Nice was placed in an enclosure on the north end of Miracles' territory on April 26. The male had been intently watching construction of the pen since its onset and was very interested when a real bird replaced the decoy that had been placed there to entice him. The interest quickly turned to anomaly when Too Nice was released from her pen. Although the female danced once for him, Miracles did not respond in kind. His primary outlet was increased aggression toward neighboring sandhill cranes. On one occasion, he knocked down a bird in flight and beat it to near death. Did he fear these birds were rivals, or was he just so confused by Too Nice's presence that he was not exhibiting normal behavior? Despite the puzzling performance, biologists watched for indications that a bond was forming between the two whoopers. By late July, they had observed merely tolerance punctuated by a single unison call. Nothing had transpired by autumn; when Too Nice made indications that she intended to migrate south with a flock of sandhills, she was recaptured and sent back to Patuxent.

Too Nice was released at Grays Lake again the following year, and Miracles resumed his war with his sandhill neighbors. Too Nice continued to ignore his antics even when he began mounding up bulrushes and cattails early in May. Drewien was hopeful that perhaps Miracles was intending to court her after all. Hopes were dashed days later when Miracles' lifeless, bloody body was found dangling from the four-stranded barbed-wire fence that bisected his territory. He had crossed that fence a thousand times; that spring day, he did not make it. He was seven years old and in prime breeding condition when he died. Too Nice continued to ignore other cross-fostered whooping cranes and was shipped back to Patuxent in October.

Nonetheless, biologists at Grays Lake continued to foster whooping crane eggs to sandhill parents for another six years. Although sandhills successfully raised and migrated 77 whooping cranes during this time, there was no evidence that they had formed pair bonds with members of their species. Even several additional force-pairing experiments failed to promote breeding. Biologists concluded that the whooping cranes had been sexually imprinted on the sandhill cranes that had raised them and were unable to identify, select and communicate with appropriate mates. By the mid-1990s, cross-fostered adult females at Grays Lake had passed through a combined 45 breeding opportunities without any selecting a mate. Only one male had become a parent, when he and his sandhill female successfully raised a hybrid chick, known as a "whoophill." Although the two species frequently associate with one another in the wild, this was the first report of cross-species pairing, obviously a consequence of improper sexual imprinting resulting from the experiment.

The Grays Lake whooping crane population peaked at 33 birds in 1985 and continued to decline throughout the 1990s; recruitment of young into the population could not keep up with the high mortality, due primarily to human-induced causes such as collisions with power lines and barbed-wire fences, hunting, and ingestion of lead shot and sinkers and other foreign objects. Many whooping cranes also died from avian cholera and tuberculosis that they had

Behavioral anomalies produced during Grays Lake cross-fostering experiments convinced conservation workers that whooping crane chicks had to be imprinted on their own species, or replicas of it.

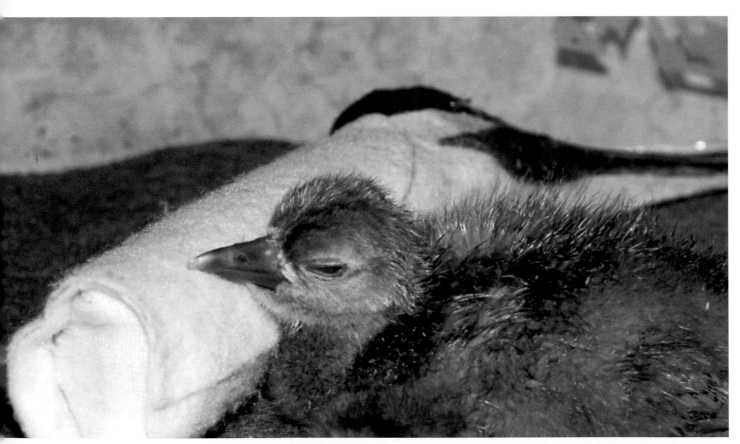

© www.operationmigration.org

contracted from snow geese during wintering. It was this risk of disease that prevented biologists from rounding up the last surviving whoopers at Grays Lake and reintroducing them into another population.

Despite untenable mortality rates in Idaho whooping cranes, the experiment had failed primarily because cross-fostered birds did not know how to be whooping cranes. Most birds, like humans, learn social behavior by watching the interactions of other members of their species. Juvenile whooping cranes learn signals associated with courtship by observing their monogamous parents strengthen pair bonds through their dances and unison calling. Behaviorists also suggest that young cranes pass through a critical sexual imprinting period between 12 and 18 weeks of age, when they learn what a prospective mate should look like. When most juvenile whooping cranes would be seeing their parents, the cross-fostered birds saw only sandhill cranes. What resulted were birds in an "identity crisis" that took its greatest toll during breeding season. Interestingly, this crisis was pervasive in its ability to produce anomalous behavior. Miracles' vitriolic actions toward sandhills cranes were perplexing. In addition, there were cases of solitary male whooping cranes building nests; one insisted on incubating an empty nest. Two other males assisted a sandhill pair with raising their chick — not typical behavior for a highly territorial, long-lived monogamous species.

In March of 1990, the decision was finally made to terminate the Grays Lake experiment. The project had lasted 15 years and had cost millions of dollars, but perhaps more valuable than time or money were the 289 whooping crane eggs expended during the study. Even at the project's onset in 1975, knowledge of improper sexual imprinting and its potential negative effects on cross-fostered species predicted that the Idaho experiment was risky, at best. Also, biometric analysis had demonstrated that the Grays Lake population would require future transfers of 30 eggs per year for 50 years if it had any hope of becoming self-sustaining. Moreover, this estimation assumed that cross-fostered birds would be contributing to their recovery through natural breeding, which of course they did not. The last whooping cranes hatched in Idaho in 1988. The 16 birds remaining in the genetically extinct flock at that time continued to travel back and forth between Grays Lake and Bosque del Apache, dwindling more in numbers each year until 2002, when pure white cranes were no longer seen standing tall among their sandhill cousins.

Despite high mortality and abnormal sexual imprinting, the Grays Lake project did produce a handful of whooping cranes that could survive in the wild and travel south in autumn. In 1993, when this Rocky Mountain population numbered only four males and four females ranging from 7 to 17 years old, biologists offered redemption for these maladjusted birds if they could serve as guide-birds to teach young captive-reared whooping cranes how and where to migrate.

Accordingly, in 1993 and 1994, three cross-fostered adults were captured during winter at Bosque del Apache NWR and transported to holding facilities near Grays Lake. At the same time, eight eggs acquired from Wood Buffalo were hatched and raised in apparent isolation by human caretakers cloaked in white costumes, sporting whooping crane hand puppets to reduce the likelihood of improper imprinting. The plan was to pen the chicks adjacent to captive wild birds to first determine if the adults would display untoward aggression toward the youngsters. If there was no evidence of disharmony, the chicks would be moved to the adults' pens, remain in their company until they fledged in late August, and then be released with them just prior to migration. Perhaps these surrogate parents would lead the young birds to New Mexico.

Unfortunately, these juveniles would not escape the bad luck that had beset the Rocky Mountain population from the beginning. Only five chicks survived their early days (four in 1993 and only one in 1994), and although the adults were docile around them, they were also inconsistently attentive. They rarely fed the young birds, and, at night, the adults appropriately selected water roosts, leaving the chicks to fend for themselves. The greatest test of their tenuous bond occurred after all the birds were released to the wild; once again, the cross-fostered whoopers would fail. When given the opportunity, they fully abandoned the young birds despite the seven months that they had been penned together. The naïve chicks attempted to abide by their adopted parents, but were ignored. Summer quickly turned to autumn and the adults flew south without their young charges.

In 1993, only three chicks remained alive after the adults departed. If they were recaptured, trucked to New Mexico and reintroduced into the wintering flock, maybe they would return to Idaho with them. Two young cranes would never get there; they died of shock and trauma during transport. The sole survivor was finally released at Bosque del Apache in November. Knowing no others, the juvenile gravitated toward any white birds — both whooping cranes and snow geese — but it was unable to form any attachments. It spent its winter alone and died of avian cholera in February.

The 1994 cohort did not fare any better. Only one chick survived its first ten days of life; two others perished of disease and one was stepped on by caretakers. Although the lone chick received some attention from the single male adult that shared its pen, it was ignored once they were released in September. It attempted to follow the adult as they foraged but, being as yet a weak flyer, it was unable to keep up as the older bird moved from field to field. Harassed by resident sandhills and ostracized by whooping cranes, the chick fed alone during the day and roosted by itself at night. It found no sanctuary even in the last moments of life. The young bird was foraging alone in a grainfield in late September when it saw its companion adult land some short distance away. Impervious to the three weeks of rebuff that it had endured, the chick scurried toward the white bird. It never got there.

A coyote hidden in the tall grass lunged and its attack inflicted mortal wounds. The nearby adult did nothing to defend it.

The Rocky Mountain guide-bird experiment had failed as absolutely as the cross-fostering project that had preceded it, due primarily to accident, disease, injury, trauma and, finally, predation. Supporters claimed some success in their ability to keep adult whoopers alive during seven months of captivity, but this would be a dubious honor. After the adults left Grays Lake bound for New Mexico something happened; all of them vanished, save one. They had traveled the same route between summer and winter for nearly a decade or two and yet this ability was lost in a few short months of confinement. Perhaps they were physically unable to travel the distance, or maybe their hold on migration was still too tenuous to weather change.

Although it appeared in these years that all human efforts to save whooping cranes from extinction had targeted captive birds, this was not at all the case. By 1990, the wild flock numbered about 150 individuals, including 32 breeding pairs. The population's growth had accelerated rapidly during the 1980s, due primarily to conservation efforts on both fronts. In Wood Buffalo National Park, CWS biologists Ernie Kuyt and Paul Goosen had spent their short northern summers counting and banding birds and switching eggs under

Captive-reared whooping crane hatchlings are shown how to feed by an adult crane puppet head operated by a cloaked conservation worker.

incubating pairs to insure that the most attentive parents would have the best opportunity to produce viable young. At Aransas, USFWS biologist Tom Stehn and refuge manager Frank Johnson had studied the cranes' non-breeding behavior and optimized their essential feeding habitats to ensure survival through the winter. The outcome of these labors was the doubling of wild whooping crane numbers in a single decade. Compared to the meager handful of birds that had passed through the bottleneck some 50 years earlier, 150 cranes seemed bountiful.

Nonetheless, this migratory flock was even now at considerable risk of annihilation from some stochastic catastrophe, such as disease or environmental disaster. Accordingly, by 1994, both Canadian and American national recovery plans had recommended that conservation efforts continue toward establishing two additional wild populations. Reintroduction of the Florida nonmigratory population was already underway by the early 1990s. On the other hand, the Rocky Mountain experiments had failed to produce a sustainable second migratory flock. Biologists, aviculturalists and crane enthusiasts, however, would not abandon their commitment to see this dream to reality. The whooping crane was far from secure, still lingering too close to the brink to just let nature take its course. Thus, the 1990s ushered in a new approach to creating a migratory flock that capitalized almost entirely on the young whooper's first instincts: the inclination to follow.

Humans have been teaching birds to follow them for millennia. Five thousand years ago, falconers trained birds of prey to accompany handlers back to their coop. Ethologist Dr. Konrad Lorenz explained this instinct to follow by the process of filial imprinting. This behavior is perhaps best expressed in precocial birds such as cranes, which are born downy and able to leave the nest shortly after hatching. At this critical age, they become rapidly and strongly attached to moving target individuals, typically parents, which they will follow without question. The evolutionary advantage of this is obvious; young birds are compelled to accompany their parents to feeding areas, shelter and generally out of harm's way until they are old enough to fend for themselves. During the 1970s and 1980s, experiments with raptors and waterfowl determined that the moving targets in question need not be living: automobiles, boats, motorcycles, all-terrain vehicles, hang gliders and airplanes would also suffice as surrogate parents for neonatal birds. So was born the idea of vehicle-led migration.

Capitalizing on these early trials was artist and ultralight pilot William Lishman, who, in association with waterfowl trainer William Carrick, taught Canada geese (*Branta canadensis*) to fly in formation behind a specially designed ultralight aircraft. Between 1993 and 1995, pilots Bill Lishman and Joe Duff successfully led 86 imprinted juvenile geese on three southbound migrations from Ontario to wintering sites in Virginia and South Carolina, a trip of over 800 miles (1,300 km). The following spring, 90 percent of these birds returned unassisted to the natal site. This amazing journey subsequently became the basis for critically acclaimed motion picture *Fly Away Home*. Lishman and Duff would later turn their sights toward crane migration, and, although they would not be the first people to follow this path, their efforts would have the greatest lasting impact on the future of whooping cranes.

Virtually all vehicle-led migration trials began with sandhill cranes, which were similar enough that experimental results could be extrapolated to predict their utility with endangered whoopers. Also, their abundance makes them readily available and reduces required red tape considerably. Consequently, between 1990 and 1992, US Geological Survey (USGS) biologist Dr. David Ellis had been conditioning sandhills to follow motorized vehicles for short distances to test their ability to fly while wearing backpack-style satellite transmitters. Three years later, he extended this experiment by using a modified US Army ambulance to lead a flock of sandhill cranes along a 400-mile (640 km) migration route through Arizona. He trained the birds to walk up a ramp into the back of the vehicle so that they could ride if they could not fly. At first, Ellis chose not to obscure his team members' human form with crane costumes because he felt it was too demanding to require crane-handlers to perpetually don cloaks and masks.

Although ingenious, the project was flawed. It was apparent that the birds suffered considerable psychological stress when they were trucked, as evidenced by imperfectly formed feathers on their back and shoulders coincident with time of transport. Rocky Mountain guide-birds had also experienced high mortality during transport, particularly among subordinate cranes that were forced to ride in close quarters with their superiors. Finally, the decision not to hide the human form produced birds that were badly imprinted on their handlers; they were attentive to little else other than their caretakers, and generally too tame and maladapted for life in the wild. The study accordingly had problems with flock cohesion, particularly near the beginning of the journey

© www.operationmigration.org

In his early experiments with vehicle-led migration, Pilot William (Bill) Lishman successfully conveyed a flock of imprinted Canada geese behind his ultralight aircraft.

when some cranes decided to fly back to camp, and after repeated golden eagle (*Aquila chrysaetos*) attacks scattered the birds. Flying at low elevation behind a ground vehicle also caused them to become overheated in the Arizona sun and put them at great risk of collision with power lines; there were 120 sets along the route, resulting in a few fatalities and many nonfatal collisions. In addition, the birds could not be led through forested areas due to the height of vegetation. All told, problems with following, forests, predation, power lines and heat exhaustion resulted in birds being trucked almost 40 percent of the route, which of course increased handling and tameness.

Changes to basic protocols in subsequent years produced birds somewhat better adapted to life in the wild; however, tameness prevailed as a problem. Furthermore, they were poor return migrants. The 1995 cohort became "nuisance birds" that required penning through the winter and refused to fly anywhere when given weekly opportunities to do so. When rereleased a year later, they moved some distance north, only to become golfers' nemeses at a local course. When removed from the area and released again, the birds flew east, not north, and set up camp in the exercise yard of a penitentiary. Perhaps they were more comfortable behind bars after all.

The 12 survivors of the 1996 cohort fared slightly better, but not much. Four birds vanished. Five were recaptured in the spring when they threatened to fly north to Nevada or Oregon with their wild compatriots. The other three flew north on schedule, heading towards the desired summering grounds, but when they encountered deep snow there, they turned around and flew back to their more temperate wintering turf. By the summer of 1999, only three birds remained from the truck-led migration experiments. They were captured and placed permanently in zoos. Obviously, whooping cranes could not be led on migration behind a ground vehicle.

Fortunately, other experiments were underway. Coincident with Ellis's truck-led migration — and likely inspired by the Lishman and Duff successes with Canada geese — rancher and pilot Kent Clegg was training captive-reared sandhill cranes to fly behind an ultralight aircraft. In previous years, Clegg had led tame cranes behind all-terrain vehicles, automobiles and trucks. He had raised the birds on his ranch near Grace, in southeastern Idaho. The proposed migration corridor would extend almost 800 miles (1,200 km) across Utah, the southwestern corner of Colorado and south down through New Mexico to Bosque del Apache National Wildlife Reserve, a route not unlike that used by Grays Lake cross-fostered whooping cranes.

Clegg chose not to use costume-rearing techniques; instead, he specifically imprinted his birds on humans so that they would become good followers. Apparently, part of the experiment was an attempt to see if this somewhat loose policy could still produce birds able to cope in the wild following release. As the chicks hatched, caretakers played recordings of human-produced crane brood calls. Clegg would later use these calls, which he made himself, to summon the birds to him. Young cranes foraged during the day, tended only by a plastic decoy of an adult, but were penned at night to avoid predation. Clegg first trained the chicks to follow him on foot and then run behind an all-terrain vehicle. Flying after an aircraft, once they fledged, just capitalized on their improper imprinting-based desire to follow humans, specifically Clegg himself.

Clegg had modified a high-wing, single-seat, open-cockpit Dragonfly ultralight. The plane was capable of flying slowly — it had a minimum flight speed of about 20 miles per hour (32 kph) — which was necessary to lead the large birds; it could stay aloft for about five hours with extra fuel. When the birds were about 20 days old, he introduced them to the aircraft by occasional flyovers, and also left the plane idling near their pen. Once they were accustomed to it, Clegg attempted to lead the cranes in flight. Initially, only one or two birds would follow their surrogate parent, but after five days of training they all fell into formation behind him. Flight lengths were gradually increased to about 25 miles (40 km) as the cranes matured and became stronger.

Despite Clegg's human-imprinting approach, only six cranes followed on the first full migration to New Mexico in 1995. A seventh bird was transported there by truck. On average, Clegg flew quite low — at mean altitudes of about 980 feet (300 m) — and marauding eagles, power lines and utility poles continued to plague the flock. Furthermore, the birds were exceedingly tame and ill-equipped to survive perils in the wild. Clegg encouraged the research birds to mingle with flocks on the wintering grounds so that they could learn more about being wild; however, these efforts were futile. The evening after arrival, the research birds in flight with the wild flock left them to join their aircraft where it was parked on the ground. The following day, they failed to follow the wild birds to the safety of their water roost. Unfortunately, Clegg and his team had fitted their birds with solar-powered transmitters and were unable to locate them after dark. They found two of them the next day; coyotes had killed them during the night.

Within three days of arrival on the wintering refuge, only two of seven research birds could be found. One missing bird was discovered later, its wing injured by a shotgun pellet. Two others turned up when hunters presented their carcasses at a crane-checking station. Although they had been requested not to shoot sandhill cranes sporting yellow legbands, such a minor detail would certainly go unnoticed by many hunters. The following spring, the survivors flew back to Idaho, following a route more typical of wild birds, and were found in April, just 30 miles (50 km) short of the Clegg ranch. The experiments continued in 1996 with a goal to increase survivability; however, Clegg still rejected the use of costumes in his captive rearing protocol. Nonetheless, he successfully trained eight sandhill cranes to follow his aircraft and they all arrived at Bosque del Apache NWR safely in October. They wintered with wild cranes and migrated north in

Most cranes led to Bosque del Apache National Wildlife Reserve during Kent Clegg's reintroduction program were killed by predators, such as coyotes, because they lacked the requisite knowledge to escape them.

spring to spend their summer in various locations in the company of their wild comrades.

In 1997, Clegg received authorization to experiment with a mixed flock of eight sandhill cranes and four whooping cranes. His protocols would remain more or less the same: no costumes. He chose, instead, to raise the two species in separate groups during sensitive development stages to reduce the likelihood of their becoming improperly imprinted. However, unlike experimental sandhill cranes of previous years, the captive whoopers would not have any adults of their own species with which to associate on the wintering ground. Who would teach them to be wild?

Migration began in mid-October; it took nine days to complete the 700-mile (1,130 km) journey; daily flight distances ranged from 16 to 115 miles (27 to 185 km) at an average speed of 32 miles per hour (52 kph). Despite some problems with eagles and collisions, 11 cranes arrived at Bosque del Apache NWR. As before, they were encouraged to forage and roost with the resident wild flock. Although survivability was somewhat improved from previous years, two whooping cranes were killed by predators over the winter. The remaining two were also at risk from predation due to their poor choice of roost site, so they were persuaded to move elsewhere. These new sites were also inappropriate, this time because of their close proximity to the refuge's public tour loop. Although the whooping cranes never interacted directly with humans, they were comfortable around them and approached them on some occasions.

The eight surviving research cranes migrated north in spring 1998. But after choosing hazardous summering sites in Wyoming, the two whooping cranes were subsequently transferred to Yellowstone National Park. This proved equally problematic when they became excessively tame from almost constant attention and adoration by park visitors, bird-watchers, photographers and school groups. The birds apparently flew south in autumn and were observed at Bosque del Apache NWR in February. The scavenged remains of one bird were found six months later in north-central Utah. The last survivor departed Grays Lake NWR on September 1 and was briefly seen near Ouray, Utah, in October. Then, it vanished.

Clegg's experiments showed real promise in their ability to induce migration in cranes. However, his renouncement of costume-rearing techniques perpetually produced birds that were too tame. Also, Clegg's research birds did not survive long enough to reach breeding age, and it was unclear what that outcome would be. It seems very likely that they were so sexually imprinted on humans, like George Archibald's Tex, that they would be unable to breed: a clear case of solving one problem just to create a second one. Yet the solution was out there.

The Eighth North American Crane Workshop was held in Albuquerque, New Mexico, in January of 2000. The gathering was sponsored by the North American Crane Working Group, a delegation of biologists, aviculturalists, land managers and other interested parties committed to the conservation of North American cranes and their habitat. They organized a workshop every three or four years to review recovery activities. At the Albuquerque meeting, a technical session had been arranged specifically to discuss the progress and results of recent crane reintroduction experiments in Idaho, Florida and Virginia. On the roster of participants were many of the programs' significant players: Clegg, Drewien, Duff, Ellis, Gee, Lewis, Lishman, Nesbitt, Stehn and Urbanek. Among them was Dr. Robert Horwich, a behaviorist and director of Community Conservation, a Wisconsin-based organization with a mission to promote global diversity and sustainable land use. Fifteen years earlier, Horwich, in association with the International Crane Foundation, had developed the pioneering technique of costume rearing to prevent behavioral anomalies in captive-bred birds. At the 2000 workshop, Horwich announced that all the knowledge required to successfully introduce a fully functional migratory whooping crane flock had been present for over 10 years. We merely had to examine the lessons learned along the way and assemble the solution.

© www.operationmigration.org

Many studies clearly demonstrated that costume-reared chicks, isolated from the human form and voice during development, were better prepared for life in the wild when released.

Horwich reiterated the primary goals of such a mission: reintroduced birds must demonstrate a healthy fear of humans and be capable of survival, migration and normal reproduction. Protocols to date had succeeded in some goals, but not all of them holistically, and indeed a great many lessons had been learned in the process. For example, it was obvious that young chicks were much better candidates for reintroduction than older birds. Yearling whooping cranes released in Florida suffered an 80 percent mortality rate in the first 15 days, and all of them eventually died. When rearing young birds, it was possible to take advantage of development stages that are more conducive to learning. The first weeks of life are critical for filial imprinting, when chicks learn to follow their parents, but by age four to ten weeks they are most influenced by specific aspects of their parents' activities. During this period, they learn to make appropriate choices regarding foraging and habitat use by watching and mimicking adult birds. Knowing how to roost in water, for instance, is fundamental for survival. Thus, it is beneficial to train young birds at their release site so that they make familiar good choices later in life. Close to fledging, young cranes learn again about following their parents: they need to stay close to avoid being lost during migration. Additional attention at this critical time could improve cohesion on a vehicle-led journey. Finally, sexual imprinting, which assures that birds choose proper mates, probably occurs between three and five months of age; interactions with same-species birds during the first winter may also reinforce correct choices.

A multitude of studies using several crane species — whooping and sandhill cranes among them — had demonstrated that costume rearing chicks in strict isolation from the human form or voice was the optimal method. Accordingly, the handler is draped in shapeless clothing that approximates adult coloration and uses a handheld puppet fashioned to resemble a parent's head to interact more directly with the chicks. Recordings of actual crane calls, not human-produced imitations, are played to elicit particular chick behavior. Post-release survival rates among strictly costume-reared cranes have been many times higher even than those that were captive-parent raised. Explanations for the obvious success of costume rearing are sometimes elusive. Perhaps allowing handlers to play parent under such unnatural conditions provides more control over both what needs to be taught to chicks and what is lacking from their education. In the wild, crane chicks rely on their parents to watch for danger throughout the autumn and winter; they do not learn to be vigilant until they gain their independence the following spring. However, chicks raised in captivity and released during their first autumn must learn how be vigilant at a much younger age.

Costume rearing also provides researchers with opportunities to control sexual imprinting during critical developmental stages. Once gone wrong, sexual imprinting cannot be reversed in older birds, particularly if they are fixed on a closely related species. This is why force-pairing experiments with cross-fostered whooping cranes did not trigger pair bonding. Fortunately, most birds are wired to imprint more easily on their own species than another — an evolutionary safeguard. Costumed handlers do not really look like whooping cranes; however, in the absence of other surrogates, costume-reared birds are more likely to default to their species. This effect can be enhanced by raising young birds with brood models and later with live birds occupying adjacent enclosures. Socializing with other conspecifics (members of their own species) helps as well.

Many reintroduction programs have produced "nuisance cranes" that do not show appropriate fear of humans. Tameness is particularly problematic near hunting or potential persecution by angry farmers or landowners.

Unfortunately, whenever research workers are lax about birds seeing or hearing them, tameness ensues; strict costume rearing certainly can reduce its likelihood. Also, if birds are not tame when they are released, they learn more about being wild when in the presence of truly wild birds. Subjecting birds to repeated mock attacks by uncostumed handlers also contributes to healthy wariness and fear.

Soft (or gentle) release into the wild has also been shown to improve survival significantly. Soft-release techniques entail enclosing young birds in a specially constructed pen for a few weeks at the reintroduction site so that they may acclimatize slowly to life in the wild. They are visited regularly by their costumed surrogate parents and are usually provided with supplemental food. Soft release techniques for cranes developed by US Fish and Wildlife biologist Dr. Richard Urbanek had been used in the reintroduction program of both endangered Mississippi sandhill cranes and Florida nonmigratory whooping cranes. Many soft-release projects boast an over 80 percent survival rate of introduced birds even one year after their lengthy migratory journey. In contrast, hard-release methods, in which adults or yearlings are introduced into the wild with no training and little preparation for their new life, typically have no survivors.

Finally, early vehicle-led migrations demonstrated the need to imprint chicks on the surrogate parent to insure faithful following, not on the vehicle that will lead the migration. Thus, costumed-reared chicks taught to follow someone in costume on the ground would also follow in the air. This paradigm would recommend that pilots abide by the same strict code of behavior as any handler, flying while wearing costumes and emitting recordings of crane brood calls as an added enticement. Technology becomes merely a tool at the disposal of surrogate parents.

Prior to 2000, the majority of crane migration trials had originated in the Rocky Mountains. However, the west presented many inherent problems, including voracious golden eagles, high elevations and mountainous terrain, power lines and hot weather, not to mention irate hunters and landowners who insisted that endangered birds in their midst trod heavily on their civil rights. Also, some ornithologists doubted that whooping cranes were ever plentiful in Idaho, Utah, Colorado and Arizona, where these researchers intended to lead them. Whooping cranes are grassland birds, and if their historical distribution did stretch this far west it was, indeed, the very edge of their existence. If a western migratory population could be established, would there be sufficient habitat remaining — given what has been modified by humans in the past 150 years — to support a sustainable number of birds? Perhaps the solution lay to the east.

In 1995, recovery teams from Canada and the U.S. formed the International Whooping Crane Recovery Team (WCRT) to provide policy recommendations to the US Fish and Wildlife Service and Canadian Wildlife Service

regarding the species' recovery programs. The team consisted of ten crane experts, five of whom were appointed by the respective wildlife services in each country. The whooping crane recognized no political boundaries; hopefully, the WCRT would not either. During the next three years, the WCRT searched North America for the optimal location in which to establish the second migratory population. Summering habitat would require extensive shallow wetlands with clean standing water less than 24 inches (60 cm) deep, sufficient food resources, not too many predators, no lead shot or other toxins and a low incidence of communicable avian disease. It was also critical that public access to restricted reintroduction areas be easily controlled. Many regions were considered, although few met all the necessary criteria. In addition, some potentially suitable areas, including parts of Saskatchewan and Manitoba, were considered too near the migration corridor of the Wood Buffalo–Aransas flock.

Wisconsin, however, showed promise. It was at the core of historical whooping crane distribution but would provide flight paths that were geographically well east of the wild flock. Also, potential habitat was available on

A costumed conservation worker leads eager young cranes on an early morning outing to the marsh.

federal, state and private lands. Perhaps equally important was the state's long history of commitment to environmental concerns and overwhelming public and political support for the crane recovery project. The WCRT enlisted biologist Dr. John Cannon to assess various sites in Wisconsin and report on their suitability. After due consideration, three potential release sites were identified: Crex Meadows State Wildlife Management Area, Horicon National Wildlife Refuge and Necedah National Wildlife Refuge.

Of course, wintering grounds were also chosen. Historically, flocks breeding in the Midwest probably wintered east of Texas, perhaps in Florida. Cannon was dispatched to survey the Gulf Coast for potential sites. After some deliberation, the WCRT recommended in September 1999 that a second migratory flock of whooping cranes be established, using ultralight aircraft to teach a migration pathway between central Wisconsin and Chassahowitzka National Wildlife Refuge, on the west coast of Florida about 65 miles (100 km) north of St. Petersburg. This reserve included about 31,000 acres of saltwater bays, mangrove islands, estuaries and brackish marshes at the mouth of the Chassahowitzka River. It was also home to the endangered West Indian manatee (*Trichechus manatus*).

At the outset, it was understood that this reintroduced flock would receive the nonessential experimental population (NEP) designation. The NEP rating is a special category under the Endangered Species Act (ESA) that would ultimately provide conservation workers with flexibility. Introducing a migratory species is risky, and past projects have seen more failures than successes. Nonessential just means that this second flock cannot be considered an absolute requirement to the survival of the species as a whole. Experimental status would help rally additional support from landowners. Migratory cranes may wander. Under the strict ESA mandate, any land inhabited by an endangered species, even privately owned land, becomes subject to harsh government regulations. Cranes from this project could potentially impact landowners in 20 states and provinces along the flyway. However, the experimental rating would guarantee that they would retain the same rights to their lands that were present before the project was initiated. In the expectation of NEP designation, every state between Wisconsin and Florida eagerly jumped aboard.

Over the years, representatives from many government, nonprofit and private organizations met regularly to contribute their knowledge, financial support and enthusiasm to this prospective reintroduction project. Their continued commitment would be essential to its success. In 2000, the Whooping Crane Eastern Partnership (WCEP), would be officially established with each of nine founding partners providing an integral piece of the puzzle. The WCRT would oversee all aspects of the project. The US Fish and Wildlife Service was federally mandated under the ESA to provide species recovery options. They would be primarily responsible for operations at the Wisconsin release site and wintering grounds in Florida. They would also coordinate state activities along the flyway, address budget concerns and provide avenues for public outreach and communication. At the state level, the Wisconsin Department of Natural Resources would offer resources for habitat assessment and manage some essential wetland areas. The USGS Patuxent Wildlife Research Center would provide most of the chicks needed for the project, as well as costume rearing at their Maryland facility before the young birds were old enough to be shipped to Wisconsin. The USGS National Wildlife Health Center offered veterinary consultation, diagnostic services, health risk assessments and avian disease expertise. The International Crane Foundation would also supply chicks. Furthermore, their global expertise with cranes and their dynamic avenue for fundraising, public education and outreach would be indispensable. The National Fish and Wildlife Foundation and the Natural Resources Foundation of Wisconsin would source additional funds through a myriad of partnerships and fundraising programs. Finally Operation Migration Inc., a nonprofit organization founded in 1994 by ultralight pilots and artists Bill Lishman and Joe Duff, would lead the birds to their new winter home.

Operation Migration was not merely some upstart organization. In the eight years prior to the formation of WCEP, Lishman, Duff and their many dedicated supporters had successfully conveyed three bird species — Canada geese, trumpeter swans (*Cygnus buccinator*) and sandhill cranes — on ultralight-led migrations. These trials were based on work pioneered by Lishman in the 1980s; however, the geographic challenges of each migratory route, juxtaposed with the idiosyncratic nature of each species, provided more than ample opportunity to refine their founding techniques. Early experiments produced birds that would follow south in autumn and return unassisted in spring but — as with other vehicle-led migration projects — fostered inappropriate degrees of tameness in released birds. Recognizing that this trait alone could doom the success of any reintroduction program — and, indeed, it had for Clegg and Ellis — Duff and retired United Airlines captain Deke Clark endeavored to reduce the birds' exposure to humans both during rearing and on wintering sites. In 1998, they conducted a study with the sole purpose of promoting wildness in captive-bred, ultralight-led sandhill cranes, which included stricter costume-rearing protocols and human-avoidance conditioning. Lishman and Duff also acknowledged that reintroduced whooping cranes would not have the benefit of wild relatives on the wintering grounds until a substantial-sized flock was established over many years. Earlier vehicle-led migrations of Rocky Mountain sandhill cranes relied heavily on the presence of wild birds to teach wildness and survival skills to newly released birds. To address these concerns, Operation Migration selected wintering sites bereft of wild individuals to determine if experimental birds could survive the winter and initiate spring migration in the absence of avian mentors.

In summer 2000, the WCRT endorsed Operation Migration to perform a "dress rehearsal" project using sandhill cranes as a surrogate experimental species to assess if their ultralight-led migration protocols could be used to reintroduce whooping cranes into their historical eastern flyway. The release site would be Necedah (pronounced *ne-see-dah*) National Wildlife Refuge; the wintering site was Chassahowitzka. Sandhill eggs were collected in early May from wild nests located in cranberry marshes surrounding Necedah and were sent to Patuxent for hatching. Although early training occurred in Maryland, at about 50 days of age the crane chicks were shipped back by private plane to Necedah to begin flight school. Field-workers in Wisconsin included Operation Migration staff and crane caretakers from Patuxent and the ICF. As the summer drew to an end, a few last-minute details required attention. Fledglings were banded and radio transmitters were attached to their legs; unlike the Rocky Mountain birds, these cranes' transmitters were powered by batteries, not tiny solar cells. Also, Lishman and crew made a preflight trip to Florida to identify 35 potential landing sites with a grass or dirt landing strip that were sufficiently isolated from human inhabitation yet adequately close to good crane habitat.

The entourage assembled at Necedah on October 3: three ultralight aircrafts, one Cessna 182, eight ground-crew vehicles and 13 sandhill cranes ready to go. Winds were calm under slightly overcast skies. The patchy ground mist posed little concern. They would fly today. At approximately 7:30 a.m. CST, three costume-clad pilots — Lishman, Duff and Clark — fired up their engines and switched on the digital recorders that emitted irresistible crane brood calls. The sandhills lifted off. Thirty-nine days and 27 stopovers later, they arrived in Florida. Eleven birds had successfully completed the longest human-led migration in history: a distance of 1,250 miles (2,000 km). Better still, nine of the ten cranes that left Florida the following spring had returned to Wisconsin by late April, many arriving within 100 yards (90 m) of their training area in Necedah. The results were encouraging.

The WCRT met in January of 2001 to evaluate results of the sandhill trials, and decided to follow though immediately with introduction of the first cohort of experimental migratory whooping cranes. The sooner, the better; the ambitious goal of WCEP's recovery plan envisioned 25 breeding pairs in an eastern migratory population of 125 individuals summering in central Wisconsin by 2020. However, the project team faced more demanding deadlines: those imposed by nature; whoopers breed in spring. Given their crane-calibrated incubation periods and growth rates, it was imperative that the colts arrived at Necedah in early July if they were expected to migrate in October. This endeavor required permits, environmental assessments and formalized approvals from numerous jurisdictions, including the Atlantic and Mississippi Flyway councils, in addition to a final ruling to proceed by the US Secretary of the Interior. Moreover, the hypothetical reintroduced population was still awaiting its experimental nonessential designation. John Christian, assistant regional director for the US Fish and Wildlife Service and co-chair of WCEP,

knew that without it, it would be impossible to get full agreement from all public and private interests. He would add, "While every person I've ever talked to likes the idea of whooping cranes, a number are not so enthusiastic about rules that protect them." Christian coordinated the red tape while WCEP accordingly selected June 30 as the drop dead date. If all paperwork had cleared their desk by that time, and if sufficient funds were in the coffer, the reintroduction would begin. Otherwise, they would have to wait until the next year. When Gale Norton, Secretary of the Interior, held a press conference to announce the federal government's decision, 10 whooping crane chicks were already forming attachments with their surrogate parents in Patuxent. The final ruling came down on June 26. The game was afoot.

The final environmental assessment of June 2001 favored Necedah National Wildlife Refuge as the initial release site, provided that dispersal patterns exhibited by cranes after release did not demonstrate any folly in this decision. Necedah NWR is located in the Great Central Wisconsin Swamp, which at 7,800 square miles (20,200 square km), is the state's largest wetland bog. The refuge would offer whooping cranes almost 44,000 acres (18,000 hectares) of wetlands and open water, although more than 92,000 acres (37,000 hectares) of suitable crane habitat would be available in the immediate vicinity. Necedah

Pilots Joe Duff (left) and Bill Lishman (right), founders of Operation Migration Inc., have successfully led four migratory species — including whooping cranes — behind their ultralight aircraft.

At Patuxent Wildlife Research Center in Laurel, Maryland, whooping crane chicks in the circle pen are encouraged to follow the aircraft as conservation workers drop mealworms from a mechanical crane head.

NWR already manages several other threatened and endangered species, including the Karner blue butterfly (*Lycaeides melissa samuelis*), eastern massassauga rattlesnake (*Sistrurus catenatus*), Blanding's turtle (*Emydoidea blandingii*), bald eagle (*Haliaeetus leucocephalus*) and timber wolf (*Canis lupus*). Also, an active forest management program is restoring the rare, and ethereally beautiful, oak savanna habitat that once thrived in Wisconsin before European colonization.

Necedah NWR was preferred over the other two potential release sites for several reasons. For one thing, its relative remoteness from large urban areas had limited the area's human population growth in recent decades; the nearby village of Necedah boasts about 900 people. Although it is rural in nature, local agricultural development had been curtailed by relatively poor land that was considered to be only marginally arable. But perhaps the greatest asset to Necedah NWR's appeal for hosting the eastern flock of whooping cranes was the people of Necedah. They embraced the bird and rallied around the project with raffles, pancake breakfasts and Midwestern hospitality. Refuge manager Larry Wargowsky would later say that this would make the greatest difference.

Prior to their arrival at Necedah, however, the "ultracranes" — as they became known — would undergo a strict rearing protocol and rigorous training program that would began even before they hatched. Unlike as in earlier reintroduction programs, WCEP recognized that wild whooping crane eggs should be left where they were laid in Wood Buffalo National Park, and recommended that their flock be produced only from eggs supplied by captive parents at

Patuxent, the ICF and other such facilities. Ten eggs comprising the Class of 2001 were hatched in Patuxent under the supervision of Dr. George Gee. Gee earned his doctorate in avian physiology in 1967, and had worked with cranes for over 30 years. His namesake Gee Whiz, son of infamous girl-crane Tex, received his worthy moniker as acknowledgment of Gee's dedication and contributions to crane recovery. This year at Patuxent would be busier than others. Prior to 2001 many eggs produced by the center's captive colony, whose 53 adult birds included 8 to 10 breeding pairs, were slated for release into the Florida nonmigratory flock. Now there would be many more mouths to feed with the second reintroduction program underway.

Both chick-rearing projects would follow Patuxent's strict costume-rearing protocols, which entirely cloaked the human form and voice and used hand-held crane-head puppets to feed and nurture the chicks. However, training cranes to faithfully follow ultralights would require further conditioning. Even before hatching, chicks listened to tape recordings of aircraft engine noises and whooping crane brood calls. When old enough to go outside, at about five days of age, they were introduced to the aircraft itself. Operation Migration pilots chose Cosmos Phase II ultralights. With their wing removed, these three-wheeled aircraft could be driven like powered tricycles, and were accordingly called "trikes." Rendered wingless, they could be positioned close to the chicks' exercise pen; periodic revving of the engine continued their acclimation.

Exercise would become a critical component to rearing healthy chicks capable of flying the great distance to Florida. In the wild, tiny whoopers get plenty of exercise scurrying after their long-legged parents as they feed. However, early captive breeding programs frequently produced chicks with leg malformations due to inactivity. Accordingly, Patuxent's program prescribed regular, daily exercise during the first month of life, when growth rates are the highest. In fair weather, they would stroll outside with their costumed caretakers; these walks frequently included an introductory lesson on marsh life. On very hot days, however, chicks would take a lengthy swim in their private pool while costumed handlers encouraged the young birds to keep moving.

When the chicks, now called colts, were fully comfortable with the appearance, sound and movement of the trike, they graduated to the "circle pen," an enclosure about 30 feet (9 m) in diameter with 2-foot (60 cm) high wire walls. The costumed pilot could taxi the aircraft around the pen's outside perimeter while colts followed from the inside. Periodically, the pilot would stop and use a puppet head mounted on an extension arm to reach into the pen, pointing out mealworms and other tasty treats to the eager young birds. At the center of the circle pen stood a smaller enclosure called the "jealousy pen." Young cranes expend considerable energy in their first weeks establishing their dominance hierarchy, or pecking order, through directed aggressive behavior. Some birds would hound their classmates so much that more submissive ones became reluctant to interact with pilot and aircraft. These problem individuals could be placed in the inner pen, where they could observe others being rewarded for "playing nice." Peace would usually ensue.

Once the cranes were tall enough to be protected by the plane's propeller guard, they were able to follow it directly as it taxied through nearby fields. Appropriate behavior would again be encouraged by mealworms and crickets and entertaining excursions to a nearby pond. Throughout the days, as the colts grew stronger, these forays would increase in length. Eventually, the aircraft would be fitted with a small dummy wing, and the colts would receive their first look at its final shape. Biologist Dan Sprague was among the many workers who played teacher to these young charges throughout the spring of 2001.

On July 10, prior to fledging, seven-week-old colts were shipped to Wisconsin in specially constructed individual travel crates that both prevented bloodshed among aggressive birds during transport and obscured the handlers' uncostumed forms from view. A small private plane was used to transfer the birds from a regional airport in Laurel, Maryland, directly to the tiny Necedah airport, reducing necessary ground travel. Each chick received a medical examination and, after being declared fit, was moved to naturally vegetated pens and adjacent training areas isolated from public view. These pens were constructed in wetlands, and were protected from insistent predators by buried wire walls, top netting and electric perimeter fencing. In addition, costumed caretakers with their watchful eyes would deter curious wildlife. There were two pen/training sites in 2001, and colts were divided more or less equally, based on their hatch date. Separating older birds from younger ones reduced injury during squabbles and allowed workers to customize the activities according to the cohort's maturity. Flight training was about to begin.

Operation Migration pilots, staff and volunteers continued flight-conditioning activities at Necedah, observing the same strict adherence to costume-rearing protocols being used at Patuxent: no human faces, bodies or voices. The cranes adapted well to their new surroundings, in large part due to their established familiarity with both handlers and aircraft. The colts' primary flight feathers were still short during these early days, so they stayed earthbound, running and hopping along behind the plane. Mealworms dropped alongside encouraged them to follow. Their feathers continued to grow throughout summer, and one day a hop and jump on outstretched wings would become the first airborne glide. A few days later, they lived for themselves evolution's 150-million-year-old gift to their kind: they flew.

Yet learning to be a crane requires more than just flight; survival necessitates both adequate food and predator avoidance. Daily activities often included marsh walks to learn about the world among the cattails and muck. Usually Dan Sprague would take them there, using his crane-head puppet — now worked with ease, like an extension of his own being — to show the colts what lay hidden beneath the water's surface. Like a doting parent, he would stand over the birds, protecting them from harm. But Sprague and his colleagues could not always be there; once released on the wintering grounds these young birds could no longer bask in the security of surrogate parents. They would be forever wild. By mid-August, Sprague and Duff knew that the colts must spend some time alone to develop the vigilance behavior that they would need to survive. So, with much consternation, they quietly sneaked away while the young birds busied themselves with their bathing, preening and curious investigations. Several hours later, they returned to retrieve them, but the birds were nowhere to be seen. Sprague and Duff surveyed the surroundings and decided to head to a tall, grassy area that the colts favored: still no whooping cranes. Anxiety levels peaked. However, as the men approached the cattails, brood calls emanating from recorders hidden in their costumes moved into audible range, and five rusty-colored heads popped up from the vegetation. The first gentle ties that had bound them had already been cut.

"Teenage" cranes, with almost entirely white plumage, feed in north-site pens at Necedah National Wildlife Refuge not long before autumn migration begins.

By late August, flight training had progressed from a few isolated flaps to two-minute loops over the marsh behind the ultralight. Occasionally, a bird would drop to the water, ever enticed by food opportunities that might be waiting there. The flight plan, however, could not tolerate this break in concentration; the cranes needed to stick with the program if they were going to be strong enough by October. Also, vehicle-led migration required diligent following. No doubt, the team with cranes in tow would pass over many irresistible wetlands en route to the wintering grounds. So to prevent this behavior from becoming habit, the "Swamp Monster" was born. Once an ultralight was airborne with a cohort of cranes-in-training, an uncostumed handler cloaked in a camouflage tarpaulin would wait in ambush in the marsh. If the birds looked like they were about to take a break, the Swamp Monster would rise up menacingly and convince them otherwise. Just another reminder to be vigilant, and to stay on target.

Toward mid-September, when meetings and media events were gearing up to announce the upcoming migration, the young cranes were being fitted with identification bands and radio-tracking transmitters. The jewelry proved quite an annoyance, and they were generally distracted by pecking and pulling at the devices on their legs. Some birds refused to fly, and just hopped and skipped as if their new gear were too heavy to carry aloft, which it was not. And although the young cranes were hooded during the fitting, they knew some-how that the big white birds were responsible for the intrusion; the resulting loss of rapport would take some effort to overcome. The weather was poor and that did not help, either. But both cranes and pilots received a reprieve from their duties for an unexpected reason. Following the tragedy of September 11, 2001, the Federal Aviation Administration placed a ban on all VFR (visual flight rules) flying. Unlike IFR (instrument flight rules) aircraft such as commercial jets, VFR planes — including the ultralights at Necedah — are not required to carry transponders that provide air traffic controllers with identification and location information. They fly cloaked, so to speak — not good in light of September 11 events.

Ever mindful of nature's schedule, Operation Migration took the opportu-nity to introduce the two cohorts of young birds to each other. Prior to this, their interactions were limited by fencing that divided their enclosures. However, cranes in the wild uphold a well-defined social dominance structure; each bird's thorough understanding of how it fits in the hierarchy ultimately reduces real aggression.

Two whooping cranes soar off the wingtip of their Operation Migration ultralight leader during the autumn journey of 2003.

© www.operationmigration.org

During migration, this is critical to maintaining both peace in the ranks and reliable following behavior. Pecking order within each cohort had been established long ago, but it was time for all the birds to sort out their differences as a group. Both cohorts were let out of their pens onto the grassy airstrip. At first, they leapt and flapped at each other, brandishing their claws and bills. A few birds chased a few others for a short while, but within minutes the commotion subsided and they flew to the marsh as a group to forage among the cattails. So far, so good.

The success of Operation Migration's efforts have been due, to a large degree, to one basic paradigm of their mission: every whooping crane is an individual, and each individual is cherished. Occasionally, this dedication called for special treatment. By late September, the birds were flying well in a single flock for increasingly longer stretches of time. Many had already learned to surf on the wake of air that spilled off the trailing edge of the aircraft's wingtips. This wake provides added lift, which reduces the flapping required to stay aloft. Nonetheless, one small crane was developing the bad habit of regularly dropping out over the marsh. Consequently, he was subjected to "abandonment training," which entailed being left alone overnight to consider his naughty ways. The next morning, he would be led back by ultralight to the rest of the flock. Perhaps if he experienced the exhilaration of "surfing the wake," a privilege usually claimed by more dominant birds, he would be compelled to follow, and, hopefully, stick closer.

Meanwhile, workers at Chassahowitzka NWR were busy preparing for their new arrivals. By early October, they had already burned over 2,370 acres (almost 1,000 hectares) of needlerush salt marsh to improve the cranes' feeding opportunities. The release pens and a new observation blind were under construction, and the ICF was making ready their video equipment, which would eventually be used in the blind to monitor the cranes' behavior. The day of departure was set for October 15.

Final preparations were also falling into place at Necedah. The migration crew would consist of three ultralights, a Cessna 182 and an extensive ground crew toting portable crane pens, food and other supplies, travel trailers for crew accommodation and tracking vehicles filled with reconnaissance equipment. Pilots Joe Duff and Deke Clark would lead the birds; Bill Lishman would fly the scout plane, which filled the critical role of pursuing any birds that chose to be uncooperative. Top-cover crew Paula and Don Lounsbury flew the Cessna. These dedicated volunteers joined the Operation Migration team in 1993, and have long watched over Canada geese, trumpeter swans, sandhill cranes, whooping cranes and airborne surrogate parents from above. Not only do they keep a lookout for trouble, they can also maintain solid communications between the ultralight pilots, ground crew and local air traffic controllers. Furthermore, their aircraft's speed allows them to scout local flying conditions ahead, seek alternate landing locations and track wayward birds. The ground crew would include USFWS crane biologists and Patuxent's Dan Sprague.

Just days before, Operation Migration office manager Heather Ray had asked Duff if he was nervous about the upcoming event. He replied, "I'd be fine if I could just get these butterflies in my stomach to fly in formation."

The butterflies would not subside in the near term. Monday, October 15, the planned departure day, dawned cold and cloudy with strong southerly winds. The journey's first leg would require about a 45-minute flight in calm conditions; this headwind would be too much for the young birds. The crew decided to stand down. The following morning, winds had diminished and swung around to the northwest, but a quick test flight proved the air aloft too choppy to proceed — another delay.

Wednesday began under clear skies and fair winds, and the sun rose to reveal a landscape covered in thick frost. The crisp snap of northern air heralded an instinctual call from the south. It was time to migrate. And so the journey, once called "the wildlife equivalent of putting a man on the moon," would begin. At 7:15 a.m. CST, three ultralights departed from Necedah NWR with eight endangered cranes flying in formation off their wingtips. Their destination, some 1,220 miles (1,960 km) in the future, would seem a world away. Yet even then, the journey had made history. The magnificent whooping crane was once more gracing the eastern flyway after an absence of over 100 years, thanks to a handful of dedicated humans who had just made the first payment toward the debt we owe this species.

In the next 49 days, the Operation Migration team and their faithful charges would be tested by all nature of difficulty: violent weather, fog, communication failure, cramped quarters, obstinate birds, long separations from family and friends, exhaustion and fire ants. And, occasionally, one can simply go astray. About a month into the journey, near Marion, Georgia, Heather Ray got lost. She was attempting to rendezvous with a set of GPS coordinates that corresponded to a prospective landing site. Her cell phone had failed, as had the paved road on which she was traveling. As she crested a hill on that small Georgia-red dirt road, she was accosted by a figure in camouflage pointing a rifle toward her windshield. She opened her window a crack, stuttered that she was lost, and asked sheepishly if he had recently seen three small planes flying with birds. The man's face softened and he broke into a broad smile, saying, "Oh yeah, I saw them fly over with the geese honking behind them a little while ago." Good enough. Rather than track back 35 miles (56 km) to join the team, Heather decided to stay in a motel near the next day's destination. Earlier that day, however, Heather had given her camera to pilot Deke Clark, asking him to take some in-flight photographs. When she later retrieved the camera, she found two blurry pictures of a birdless sky and one of Clark and Duff with former President Jimmy Carter. The cranes had brought them all together — the planes, the President, the people — in an irresistible adventure, and the day-to-day difficulties did not matter much. As Dan Sprague would later write, "they were nothing but small bumps in a big road on a very important journey."

The entourage arrived at their final destination, a remote coastal island on Chassahowitzka refuge, on December 5. The 1.2-acre (0.5 hectare) empty pen, where these soft-released birds would spend their first southern month, was lying in wait. The last tricky maneuver to accomplish was the "whooper air-drop." As the island offered no appropriate landing strip, the pilots would fool the small flock into thinking that everyone was landing, when job one was just getting the cranes down. Sprague had already arrived by airboat and was wait-ing in costume with loudspeaker in hand. Duff and birds in tow made a low pass over the pen. On final approach, he switched off his crane-call recording as Sprague turned his on, and then Duff climbed sharp and fast before the birds could follow. Unable to recover lost altitude, they were compelled to land. Thus, Operation Migration delivered seven ultracranes safely to their new winter home; only one bird had succumbed to injury en route.

The 2001 migration required 35.8 hours of flight time spread over 50 days; bad weather grounded the team for 22 days. The longest flight en route was over two hours, the shortest 38 minutes. One day, they covered over 95 miles (152 km) in a tailwind. In the spring, these birds would hopefully retrace this distance to Necedah in a week or 10 days. From now on, they would have the advantage of all wild cranes — they could effortlessly ride the thermals on their outstretched

Young cranes follow their surrogate parent dutifully over hill and dale, and cities as well, en route to their Florida winter home.

wings. Surely, they would know how. Two weeks before, the 2000 experimental sandhill cranes had left Necedah heading toward Florida. They were seen soaring gracefully in wide circles on departure, and one day later arrived at Jasper Pulaski Fish and Wildlife Area, almost 300 miles (480 km) to the south.

Winter conditions at Chassahowitzka were not ideal. The abrasive needlerush salt marsh had grown back by February, and local tide fluctuations were such that water levels were often too high or too low to provide appropriate roosting options. Sadly, bobcats killed two cranes early in winter. Both cats were trapped and removed from the area, and no others were seen until spring. Refuge workers made a note to initiate trapping in autumn before the cranes flew in from the north. Nonetheless, the birds were acting more or less like cranes. They gorged on blue crabs, snails and shrimp during the day and sought available water-roosting habitat at night.

Moreover, on April 10 they began moving north, riding the crest of a substantial tailwind. The tracking team on the ground, which included USFWS's Richard Urbanek and ICF's Anne Lacey, began the chase. All the birds sported radio-tracking devices and two of them also bore satellite transmitters, which would allow NASA to pinpoint the birds' location if they moved out of ground-tracking range. They were traveling fast. Despite being grounded for two days by poor weather, they were in Johnson County, Indiana, by April 15, a distance of 866 miles (1,386 km). Three days later, all five birds crossed the Wisconsin border. Urbanek followed a flock of four north through the state. Everyone held their breath when the cranes threatened to miss their target, and then, in an instant, they veered west and headed straight for Necedah. They made landfall less than a half-mile from where they had spent the previous summer.

The rich waters, salt marshes and savanna of Chassahowitzka National Wildlife Refuge in north-western Florida were chosen as the winter destination for Necedah's migratory flock of ultra-cranes.

The fifth bird, a female, was still biding her time in south-central Wisconsin. On May 2, she was foraging in a cornfield with three companion sandhill cranes when two coyotes approached with stealth. Moments later, the birds exploded into flight, with Number 7 ultracrane leading the others to safety. The following evening when Urbanek made his nightly roost check at Necedah, he was startled by a whooping crane standing near the training pens' entry door. It was Number 7, fashionably late. The Operation Migration team breathed a sigh of relief, broke out the champagne and sharpened their pencils. Eighteen eggs had already been allocated for the 2002 flock, and the colts would be arriving in just three months. There was no time to lose.

The following years would see continued success for the eastern migratory whooping crane reintroduction project, with the release of 16 ultracranes in 2002, 16 in 2003 and 14 in 2004. By April of 2005, 45 birds were winging north in a well-defined migration corridor between Florida and Wisconsin. Post-release survival was excellent, about 85 to 90 percent, even after their lengthy spring migration. The years had not been without trial, however. The Operation Migration team and other WCEP members were devastated when pilot Deke Clark suffered a severe stroke on January 21, 2002. Although his rehabilitation was encouraging, he would not fly among the cranes again. His enthusiasm for the adventure has never waned, however, and he remains an inspiration to those who followed in his footsteps.

Experienced ultralight pilot Brooke Pennypacker, joined the flight team in 2002 to fill the empty cockpit. A fourth ultralight, piloted by Richard van Heuvelen, was added to help tend the enormous flock of cranes on the wing that year. Many improvements were made on both summer training grounds and wintering grounds, particularly to night-roosting opportunities. During the autumn, 95 tons of broken shells were transported in 200 loads by two helicopters to create sloped roosting habitat at the salt marsh release pen at Chassahowitzka NWR. Maryland blue crabs were shipped from Patuxent to young birds-in-training in Wisconsin to whet their appetite for wild foods.

Migrating with such a large flock — wild whoopers typically travel in twos and threes — posed its own set of new challenges, and pilots frequently resorted to in-flight maneuvers that looked more like "crane corralling" than migration. Nonetheless, it would have been easy for a pilot to forget previous aggravations on a day such as November 12, when 15 rusty-white whooping cranes — as many as had comprised the global migratory population some 60 years earlier — surfed off the plane's wingtip for almost two hours and 90 miles (145 km) of this historic journey. Several days later, the entourage would be joined briefly over Tennessee by two more cranes, southbound migrants from the 2001 brood seeking their own way to Florida in autumn. That year's flock would journey 49 days, but all 16 birds would arrive safely.

Crane corralling reached its pinnacle as an art form during the 2003 passage. Although migration-route geography poses several challenges, none can compare with the Walden Ridge, which spans Cumberland and Meigs counties in the Appalachian Mountains of Tennessee. The day-trip needed to cross the 2,800-foot (850 m) high ridge is 47 miles (75 km) long, and requires a steep 2,500-foot (760 m) aerial switchback climb within 5 miles (8 km) of the start. Leading birds up the first 500 feet (150 m) is relatively straightforward. However, when they encounter headwinds at this altitude — and are still in sight of their temporary pens on the ground — they have a tendency to change their minds. On this occasion, Pennypacker was first to brave "the Beast" with the flock of 16 cranes. When the birds reached the buffeting winds, they began to break off. The other two pilots circled behind, trying to pick up and push up any birds that chose to form off their wingtips. The ground crew rallied the Swamp Monster, carried south for this distinct purpose, and honked truck horns to discourage the cranes from heading back. Two birds were sticking close to van Heuvelen's wing at 3,000 feet (900 m), so Duff gave him the go-ahead; he quickly disappeared over the ridge with his companions in tow. Fourteen still to go. By this time, Pennypacker had reacquired the main flock; however, with each 100 feet (30 m) of altitude gained, he would lose a few birds. Duff flew in to pick up the stragglers but soon found that they would only follow him faithfully when his crisscross rising path was oriented in pen direction. Four could not be encouraged any higher, and retreated back to their handlers on the ground. Five miles to the west, ten cranes off Pennypacker's wing also changed their mind. He and Duff managed to intercept them as they flew east, but lost four again when they dropped to treetop elevation. The ground crew was notified, and the Lounsburys, flying top cover 1,000 feet (300 m) above, tracked the itinerant birds to ensure that they arrived there safely. After 47 minutes in the air, Duff and Pennypacker finally had adequate altitude to cross the ridge with five cranes between them.

Meanwhile, three of the four cranes being tracked by top cover had not returned to their pen. Instead, they were gaining altitude slowly while flying just above the trees. They were navigating a distinctive southern path that could only be interpreted as a search pattern, perhaps hoping to reconnect with their lost surrogate parents. Nevertheless, the cranes were crossing the ridge by themselves. Unfortunately, they could not have known about the five-mile airspace restriction surrounding the nearby Watts Bar nuclear power plant. Suddenly, two F-16 fighter jets screamed past, scattering the small flock in all directions. The top-cover pilots, thinking that the military aircraft were on an intercept flight, radioed authorities to verify and were relieved to hear that the jets were merely on maneuvers. But as the cranes continued to circle closer to the power plant, the Lounsburys knew that they would have to break off the chase.

Tracking vehicles were dispatched on the ground to search for the wayward cranes. Fortunately, Anne Lacey was in Tennessee following the southbound journeys of the 2001 and 2002 cohorts, so her team in their Cessna were called in for assistance. The birds were found circling 20 miles (32 km) east of Hiwassee Wildlife Refuge, where the Operation Migration team had assembled overnight pens. The ground crew searched frantically for a place where costumed handlers could call down the remaining three cranes, shifting from one potential site to the next, jumping frantically in and out of their truck as the birds moved above them. In the town of Athens, they sought out a baseball diamond. The groundskeeper was startled by the sudden arrival of two alien creatures dressed in baggy white robes, waving radio antennas skyward and blaring unusual noises from loudspeakers. Three birds were eventually brought down in an isolated field; they had been aloft for 5 hours and 22 minutes. The last crane was found relaxing in a pond not far from the day's starting point. It was crated and trucked to Hiwassee along with its other reluctant comrades. The Beast had been conquered once again, but the ultimate goal was still more than 500 miles (800 km) to the south. The following morning at 6:15 a.m., the migration team would begin again.

Sixteen cranes joined 19 others at Chassahowitzka NWR on December 8, 2003. Although this trip had taken 54 days, it boasted a single day's flight that ranked as the longest ever made by an ultralight aircraft leading a flock of cranes: a phenomenal 200 miles (322 km) in 3 hours and 4 minutes. The next year, 14 more ultracranes set down at Chassahowitzka. Elsewhere in the south, the news was also good. The Wood Buffalo–Aransas flock numbered a record 215 birds, including 33 young and at least 67 breeding pairs. The nonmigratory flock in central Florida had also clawed back from the edge of existence and now totaled no less than 66 birds.

Another much talked-about release program, Direct Autumn Release (DAR), was initiated vicariously in 2004 when a young crane, known as Number 18-04, unknowingly became the guinea pig. DAR had been considered for several years as an addition to the primary reintroduction project of aircraft-led migration. It was based on the premise that fledglings released coincident with the autumn premigratory staging period would follow experienced birds to the wintering grounds and thus would learn the route. DAR had the benefit of being much less labor-intensive than vehicle-led migration, and could be used to supplement the introduced population once there was sufficient representation of migratory adults. The program also provided a safety net by which young birds deemed somehow unsuitable for aircraft-led reintroduction could still participate; these partially conditioned birds would be otherwise relegated to a captive flock. Earlier experiments using sandhill cranes by USFW biologist Dr. Richard Urbanek supported the potential success of DAR with whooping cranes.

Crane Number 18-04 had hatched later than the other ultracranes, and he was often distracted during training. In addition, he had experienced a primary feather growth problem due to an earlier medical condition. This little bird needed more time than the aircraft-led project could offer. Consequently,

Number 18-04 was released in late October into a flock of adult whooping cranes at Necedah NWR. He successfully followed the adults to Chassahowitzka in a 57-day adventure that included a 5-week layover in Hiwassee Wildlife Refuge in Tennessee. The success of Number 18-04's southbound journey endorsed the proposal to begin the DAR program in earnest during summer 2005.

Preparation for the 2005 ultracrane cohort began in early spring not long after previous years' birds began their now habitual passage between Florida and Wisconsin; the WCRT had allocated 24 eggs from both Patuxent and the ICF. After only five years, WCEP danced tantalizingly close to the halfway point in their lofty goal of 125 birds in the eastern migratory flock. In addition, the oldest ultracranes would be four years old that spring, just about the right age to start thinking about forming bonds. A few pairs had defended territories at Necedah the previous summer. But successful breeding was, perhaps, the toughest stumbling block in any recovery project. Were the ultracranes wild enough to choose appropriate mates and produce viable young?

Four Operation Migration aircraft would take the birds to Florida in 2005 with Duff, Pennypacker, van Heuvelen and new pilot Chris Gullikson at the controls. Changes were also made to the flight itinerary to alleviate recurring behavioral problems with the birds. One was a tendency for some individuals to break off from the aircraft shortly after take-off on the first day of migration, a setback that was decidedly irritating to all involved. Although this day has considerable significance to human members of the team, it appears little different from any other day in training for the birds. Sometimes, they just do not feel like going anywhere but home. Once the young birds were over unfamiliar turf, however, they tended to stick to their surrogate parents like glue. Hence, the team devised a "proximity strategy" whereby the first leg of their southbound journey was only a few miles from Necedah; any stubborn birds would be relatively easy to fetch. In previous years, they had traveled 23 miles (37 km) or more that day, a trip that also included a demanding pass over Interstate 90/94. Perhaps Day One of the 2005 migration would see a smoother start.

Changes were also made to the expedition terminus. Bird migration only appears to work like clockwork. In reality, there is considerable difference in when birds initiate their journey, the greatest variable, of course, being weather. In 2004, a lack of arctic weather systems bearing strong northerly tailwinds delayed the departure of older ultracranes in the eastern flyway. Many of them did not leave Wisconsin until well into November, with a few delaying until mid-December. When these "higher-ranking" birds arrived on the wintering grounds, the new cohort was already there, feeding and roosting in their space. The aggression that the experienced birds directed toward the younger ones caused considerable havoc, and increased their susceptibility to predation. The team built a small enclosure to harbor the younger birds and prevent older ones from taking their food. It was a good temporary solution; however, the arrangement required the young birds to be penned more than was deemed

beneficial, and reduced their time acclimating to their wild winter home.

The plan was to provide a transient station for the new cohort at Halpata Tastanaki Preserve, 26 miles (42 km) northeast of Chassahowitzka NWR, just for a few weeks to let older migrants get established. Once they had passed through, the ultralights could be called in again to fly the new birds the rest of the way. Of course, this would require the pilots to don their costumes again in January, again setting their own lives aside for the cranes' benefit. This was nothing new. This type of dedication blended with flexibility had essentially guaranteed success of the eastern migratory flock from the outset.

Spring had sprung at Necedah NWR early in April 2005. Within a few days of the cranes' return from the south, six pairs were dancing and unison calling among the cattails. Another pair declared their claim to space just to the east, at Meadow Valley State Wildlife Area. By mid-month, five pairs were building nests. Refuge workers and WCEP team members held their breath in anticipation. Perhaps 2005 would be the year that ultracranes led their own rusty juveniles along the route that they had been taught. The following week, when a single egg was laid in each of two ultracrane nests at Necedah, refuge manager Larry Wargowsky prepared to pass out cigars. The celebration would have to wait for another year, however, when predators took both eggs. Although heart breaking, it was not unusual for such young parents to lose their first clutches. Cranes learn from their mistakes, nevertheless. Maybe they would be successful next year.

Meanwhile, long hours of hard work began in the Direct Autumn Release program with Richard Urbanek at the helm. Five whooping crane eggs were allocated to the program that first year. When the chicks were about 30 days old, they were transferred to an isolated research facility at Necedah NWR, where they were raised under strict costume-rearing protocols. Their handlers led them outside each day for exercise and to explore the neighboring fields and marshlands, and as they grew their education included lessons on foraging and water roosting. By late August, the four DAR birds (one had succumbed early to injury), now about 70 days old, made their first flight. The birds matured rapidly, and in early September the fledglings were old enough to roam their world freely, with regular inspections by their costumed "parents." At night, they were safely locked up in a top-netted outside wet pen that enclosed suitable water-roosting habitat. As autumn began, the fledglings were exploring their world in large aerial loops over fields and above the treetops. They were regularly visited by five inquisitive adult ultracranes and were well acquainted with the sandhill couple that lived nearby.

By serendipity, whooping crane Number 18-04 became the first Direct Autumn Release bird when he successfully followed a flock of southbound migrating cranes to Chassahowitzka National Wildlife Refuge.

October waned and the young birds were fitted with colored leg bands and radio-tracking transmitters, then fully released in the wild amid the adult cranes. Despite their inexperience, they usually foraged and roosted in appropriate places. The four DAR birds shared each other's company most of the time, and were frequently seen together in the Necedah wetlands through the third week of November.

Meanwhile, the Operation Migration team was already en route to Florida. They had left the refuge on October 14 with 20 young birds off their wingtips. The proximity strategy and additional aircraft proved invaluable that first day, particularly with such a large group of birds to manage. By the time the DAR birds were released, however, they were only in Kentucky, after almost 40 grueling days plagued by fog, strong headwinds and driving rain. Many southern stopovers had received so much precipitation that standing water remained in fields that had been dry in previous years. On several occasions, their 20 young charges were reluctant to leave such interesting and hospitable habitat. Yet everyone endured and the entourage continued south.

Back at Necedah, a major cold front passed through late on November 23 and the DAR birds arose to clear skies and a strong tailwind — optimal migration conditions. Accordingly, at 10:33 a.m. CST, they departed in the company of more than 50 sandhill cranes. The young whoopers were moving fast, and had outdistanced ground trackers by the time they reached Illinois. Trackers picked up the satellite signal of a female named Jumblies (DAR 28-05) that night. She had flown 455 miles (730 km) on her first day of migration and was overnighting near Louisville, Kentucky. The following day, the other three DAR birds were tracked to Hiwassee Wildlife Refuge in Tennessee. Jumblies stayed in Kentucky for several days but then continued south to Rhea County, just across the river from Hiwassee, where she would later reunite with Poe (DAR female 28-05). They would stay there together throughout the winter. Maya (DAR female 33-05) and Waldo (DAR male 32-05) continued south to Florida in early December, where they wintered among sandhill flocks in Alachua County and on the Kissimmee Prairie in Osceola County, not far from the reintroduced nonmigratory whooping crane flock.

Late in February, the DAR birds began to move. Jumblies and Poe left Hiwassee together on the 26th, but split a few days later when they chose different sandhill flocks as traveling companions. But despite lengthy stopovers in Indiana, both birds arrived in Necedah NWR during the last week of April. Waldo started his journey north in late March. He, too, homed to Necedah and arrived on May 18 with two sandhill cranes. Soon after, he switched his allegiance and summered with two ultracranes. Maya had her own ideas. She left Florida in late March and had moved through northern Georgia, Tennessee and Kentucky and into Indiana by April 1. Her roost that night was only 8 miles (13 km) north of a 2005 ultracrane flock of 14. Strong western winds pushed her east and she set down in Barry County in southwestern Michigan.

The DAR team attempted to relocate Maya to Necedah, but they were unable to capture her among her many sandhill companions; at least, she was not alone. Notwithstanding Maya's wandering ways, the first-year DAR results were favorable. Four chicks were allocated to become the autumn 2006 DAR cohort.

By late March, five years of ultracranes were winging their way north, and the wintering grounds were nearly empty. There were now 64 cranes traveling the corridor between Florida and Wisconsin. Since 2001, 76 birds had been introduced into the eastern flyway by WCEP, and survival rates well over 80 percent were certainly something to cheer about. Three of the original unwritten program goals had been achieved. They feared humans, they had survived and they had migrated — now, would they breed?

Seven ultracrane couples had formed in Wisconsin the previous year, which resulted in two eggs being laid, then subsequently lost to predators. Early signs indicated that perhaps twice as many birds would pair up in spring of 2006. Everyone at WCEP kept their primaries crossed. By mid-April, the local tracking team was spreading the good word: five pairs were incubating eggs. The cigars seemed imminent. Spirits flagged a few days later when two clutches were lost to predators. Despair deepened on April 20, when two other prospective parents were seen one evening foraging together some distance from where their nest had been, indicating that they,

Whooping cranes in training fly in formation over the luxuriant marshes of Necedah National Wildlife Refuge in south-central Wisconsin.

too, had lost their eggs. A check of the nest site the following morning confirmed this. Only two pairs remained.

On April 24, ultracranes 213 and 218 left their eggs unattended for two hours. They had been diligently incubating their clutch earlier in the month but, for reasons unknown, relinquished their responsibilities almost three weeks later. Fearing that the unprotected eggs would also fall prey to predation, they were collected and sent to Patuxent for incubation. Workers placed a dummy egg in the nest, just in case the adults changed their mind and wished to continue their vigil. Although no amount of incubation could hatch a plaster egg, a little more experience at being a doting parent would not hurt.

The bad news was delivered on April 30. The last of five incubating pairs — these two occupying the Monroe County flowage territory — had also lost their clutch. It was always possible that any of these pairs could nest again; it was still early in the season. Nonetheless, everyone went back to their daily routines, tracking the activities of previous cohorts and preparing for the next. Spirits lifted somewhat in early May when Mark Nipper at Patuxent announced that the two eggs removed from the Necedah nest had hatched. Although this was not natural reproduction, it was reproduction, nonetheless. John Christian would later comment that even these baby steps represented

Two chicks born to ultracrane parents in June of 2006 on East Rynearson Pond were the first whooping cranes to be hatched in the wilds of Midwestern America in over a century.

"a great leap forward in whooping crane recovery." Considering the state of populations some 65 years ago, he was probably right.

Progress reports throughout May were filled with news from Patuxent. Prospective members of the 2006 ultracrane cohort were cracking out like gangbusters. Amid the birth announcements was a simple note regarding ultracranes 211 and 217, the Necedah pair that had lost their clutch to predators on April 21. By late May, they were nesting again. They were reported to be still incubating on June 13, and on the 20; Richard Urbanek delivered the news on June 23. The parents were off the nest, and were observed stamping down the vegetation in its vicinity. Their behavior suggested that they were fetching food. Something had indeed happened at East Rynearson Pond dike. Urbanek took a closer look. Not one, but two, whooping crane chicks had hatched in the wilds of Midwestern America for the first time in over 100 years. Dedicated WCEP workers felt like grandparents that day, and Larry Wargowsky distributed the long-awaited cigars. Christian hailed the event as "the start of a new generation of wild things, and a symbol for restoring our wild places."

Perhaps everyone also gave a small sigh of relief. The eastern flyway reintroduction program — "six long years of dedication, vision, and believing that it could happen" — had been remarkably successful; however, the true test of its value was self-sustainability. This is how wild populations endured, and where other reintroduction programs had ultimately failed. However, young whooping cranes have a long road to travel from hatching to fledging, and, although their parents display the greatest concern for their welfare, the perils of their childhood are many. These four birds, christened the First Family, would capture the hearts and minds of two nations. The answers to two essential questions would be revealed in future months: could the chicks survive their first summer, and would their parents lead them south in autumn? For only then would the circle of life, started so many years ago, be complete.

Chapter 5

Summing Up

Sixty-five years ago, the future of the whooping crane looked bleak. A scant 15 birds migrated semiannually between coastal Texas and some yet-to-be-discovered northern breeding grounds, and six more clung to a stagnant existence near White Lake, Louisiana. This handful of individuals remained merely the tragic remnants of a once-bountiful species that nested throughout the North American grasslands. They surely seemed destined to go the way of the dinosaur, or the dodo. Even best-case scenarios abandoned whooping cranes to the annals of protective custody, like California condors (*Gymnogyps californianus*) and black-footed ferrets (*Mustela nigripes*), whose entire being existed for a time only in captivity. Yet, whooping cranes may have beaten the odds; they are coming back.

Why this species, in particular, has been salvaged from oblivion remains a mystery. There are, indeed, thousands of others that could claim the same need for human intervention. Perhaps it is the sheer magnificence of the whooping crane that has fueled our enthusiasm; maybe their tenacity has simply mitigated their recovery. Throughout our fumbled attempts to restore what we took by the plow and the gun, the birds stood fast, tirelessly leading their young though the perils that lay between winter and summer.

Two years ago, Brian Johns (Canadian Wildlife Service) and Tom Stehn (US Fish and Wildlife Service) compiled the 2005 Draft Revised International Recovery Plan with the assistance of a myriad of other workers from office, field and laboratory. The document provides guided recommendations that, if fulfilled, would allow authorities to downlist the whooping crane from endangered to threatened. A rare honor to be sure; despite efforts of dedicated conservationists, many endangered species depart their status through either extinction or duress from self-interested lobbyists. Nonetheless, the whooping crane recovery plan outlines two primary objectives with measurable numerical criteria that would represent considerable success in the species' restoration, provided that multiple threats to its continued existence — habitat loss, hunting, urbanization, water management projects, industrial development, inbreeding depression, disease and climate change — are, somehow, adequately reduced or removed.

Objective One requires the establishment and maintenance of wild, genetically stable, self-sustaining populations of sufficient size to weather stochastic (randomly determined) environmental crises. An estimated minimum of 40 productive breeding pairs among no fewer than 160 adults in the migratory Wood Buffalo–Aransas population would meet this criterion, given that two additional self-sustaining and geographically isolated

A lone whooping crane stands tall among gray-plumaged sandhill cranes at Bosque del Apache National Wildlife Refuge in New Mexico.

populations of at least 100 individuals each, including 25 breeding pairs, could be reintroduced. If these populations prove to be nonviable, however, then the primary migratory population would have to exceed 1,000 individuals, comprising 250 productive pairs, in order to be considered resistant to catastrophe. The plan further recommends that downlisting would be advisable only if these target goals could be maintained for at least a decade. The second objective outlines the need for a genetically stable captive population as a hedge against extinction and to provide a source of birds for release into the wild. These captive birds would represent the gene bank for the entire species. Consequently, they would be required to preserve 90 percent of the existing genetic material inherent to extant whooping cranes for a century. The recovery plan suggests that 21 productive pairs among 153 captive birds would retain this safeguard.

The document makes few recommendations regarding removal of the species entirely from the List of Threatened and Endangered Species. Despite recent successes in whooping crane recovery, they nonetheless remain critically endangered; downlisting by the year 2035 seems an adequately lofty goal for now. Furthermore, it is difficult to predict the minimum viable population size needed to ensure existence in perpetuity. Healthy whooping crane populations have eluded the North American continent since European colonization. Some experts suggest that 7,000 individuals would be sufficient, but until biologists have the opportunity to assess long-term survival potential of existing fragment populations, it will be difficult to know for sure. Nevertheless, in September 2006, whooping cranes achieved a significant milestone — they totaled 500 individuals globally, with 354 wild birds and 146 more in captive colonies. Certainly, that is cause for some celebration.

The Wood Buffalo–Aransas flock has been the cornerstone of the species since recorded history. Indeed, genetic analyses have recently determined that the apparent fragment population that existed in Louisiana prior to 1950 was merely a secondarily isolated piece of the original continental distribution, which included the Wood Buffalo birds at its northern extreme, not unlike two islands that were once part of the mainland until rising water levels separated them. Ongoing efforts to increase the size of this flock — such as summer, winter and migratory stopover habitat management, supervising productivity of breeding pairs, elective manipulation of eggs and migration monitoring — have been encouraging. April 2006 population estimates placed this flock at 214 birds, including 25 juveniles and 69 breeding pairs. As numbers increase, however, acquiring reliable estimates has become progressively time consuming. Tom Stehn, USFWS biologist and co-chair of the International Whooping Crane Recovery Team, once declared:

I sometimes think that every white pelican, great and snowy egret for miles around flies to Aransas every week just to get counted. Throw in the occasional piece of white styrofoam trash washed up into the marsh along with white refuge boundary signs, and our eyes have much to sort through to find all the whooping cranes.

Counting difficulties aside, the wild flock may have a promising future. Since 1938, the average annual rate of growth has been about 4 percent. Accordingly, this flock is projected to reach 500 individuals by the year 2020, with less than 1 percent chance of extinction in the next century. Under these conditions, the first objective of the draft recovery plan could be considered to have been achieved. There exists one significant caveat to this optimism, however. These figures are based on the assumption that mortality rates — due to negative factors such as diminishing habitat or environmental catastrophe — stay more or less the same as they are currently. The healthy population increases enjoyed by the Wood Buffalo–Aransas whoopers in the last half of the 20th century have been generally due to declining mortality rates, not increased reproductive success. The greatest impact has clearly been an overall reduction in the number of whooping cranes killed by hunters and irate landowners; however, any change in mortality rates, even an oil spill on the Texas Gulf Coast, could relegate this wild flock to peril once again.

The outlook for the reintroduced nonmigratory flock in central Florida may not be so rosy. Initiated in 1993, this population has been plagued with high mortality rates, particularly due to bobcat and alligator predation within the first months of release. For many years, these challenges were exacerbated both by extended drought, which reduced standing water to hazardously low levels, and emerging diseases such as disseminated visceral coccidiosis, a chronic, debilitating infection that affects the birds' respiratory and digestive tracts. As a result, only 87 individuals from 268 captive-reared released birds survived after one decade of the program.

Early viability assessment studies predicted that the Florida flock faced a relatively high risk of extinction unless human intervention could reduce adult mortality and increase potential reproductive output. Other work suggested that chicks reared by captive parents at Patuxent fared somewhat better because they congregated in larger flocks and demonstrated greater vigilance once released; these are both effective antipredator strategies. Unfortunately, there are not enough captive parents to provide for the program; crane parents typically raise only one chick at a time, while costumed humans can tend many.

Prospects for self-sustainability improved somewhat when the Florida nonmigratory flock became productive. In 2002, Lucky held the honorable position of first chick born to reintroduced parents; two young fledged the following year. The 2004 breeding season also began with optimism; a record 12 pairs nesting in Florida produced 22 eggs among them. Celebrations were short-lived, however, when only one chick survived long enough to fledge. In 2005, the lone chick that hatched died at six days of age. Many other eggs laid during these years proved to be infertile. Early in 2005, the WCRT decided to suspend further releases into the Florida nonmigratory flock until its systemic problems with high adult mortality — which still averages about 15 percent annually — and desperately low reproductive success could be more completely evaluated. The tenacity of whooping cranes may be called to duty once again. At present, the future of the Florida nonmigratory flock can only be found in the collective memory of these birds, provided that they can recover their ancestors' ability to successfully raise their young to fledging.

Since reintroduction efforts began during the 1970s, captive whooping crane colonies have become a cornerstone in the preservation of precious genetic diversity and in providing chicks for release efforts. Prior to this time, the captive propagation program, such as it was, had fallen onto the narrow scapulars of only four birds — two males, Crip and Pete, and two females, Josephine and Rosie. The venerable, long-suffering Josephine lived in captivity for 24 years and produced 13 chicks. Four of her offspring lived more than a decade but, ill managed, they left no survivors. On the other hand, Rosie and Crip produced Tex, the crane with a penchant for dark-haired men. With her surrogate mate, George Archibald, at her side, Tex begat Gee Whiz, who subsequently fathered and grandfathered generations of cranes, including birds released into the nonmigratory Florida flock and ultralight-led migrants in Wisconsin.

Whooping cranes proved more difficult to raise in captivity than sandhills, but, nonetheless, workers persisted, often with little more than a bird or two in the beginning. Canus, who was captured at Wood Buffalo in 1964 as a chick with a broken wing, was the first member of the Patuxent flock. Canus died in 2003 at the ripe old age of 39 years, but his legacy lives on in 186 descendants that he contributed to the recovery of his species. Perhaps his well-chosen name is truly symbolic of these two nations' cooperative efforts to bring the whooping crane back from extinction.

Captive colonies have since grown to 32 breeding pairs among 135 individuals at nine facilities, including Patuxent Wildlife Research Center (53 birds), the International Crane Foundation (32), Devonian Wildlife Conservation Center in Calgary (17), the San Antonio Zoo (8), Freeport-McMoRan Audubon Species Survival Center in New Orleans (8), the Calgary Zoo (2), Lowry Park Zoo in Tampa (2), the Audubon Park Zoo (2) and Homosassa Springs Wildlife State Park in Florida (1). In addition to housing and rearing captive cranes, these facilities also participate in disease research, vaccine development and genetic typing. It is apparent that whooping cranes passed through a severe genetic bottleneck around 1940, when only a few breeding pairs remained. Evidence of extensive inbreeding, including congenital deformities of the feet and legs, are regularly observed. Poor reproductive effort and infertile eggs, as exhibited by the reintroduced Florida flock, may also result from inbreeding depression. Consequently, genetic management — which basically serves to distribute what little variation exists to its best advantage — has recently become an important part of species recovery programs, and it will continue to be so in future. Nonetheless, the WCRT draft recovery target goal for captive colonies is just a hand's reach away.

Operation Migration and other WCEP partners have basked in unprecedented success in reintroducing an eastern migratory flock of whooping cranes since 2001. By autumn 2006, more than 60 adult cranes were migrating in a narrow corridor between Wisconsin and the Florida Gulf Coast. With a few exceptions — generally dispersing females — they were all faithfully returning each spring to the Necedah NWR environs to reclaim their summer turf. After only five years, WCEP is more than halfway to reaching the 2005 draft recovery plan goal of 100 individuals in the eastern migratory flock. Assuming projected losses of 10 percent annually due to adult mortality, it would take 10 years with at least 20 birds reintroduced each year to achieve this goal. The 2006 migration cohort represents the project's sixth year.

© www.operationmigration.org

The offspring of two ultracranes, Number 602 was hatched in captivity when her parents' nest at Necedah National Wildlife Refuge was threatened by predation.

The recovery plan also stipulates a requirement for 25 productive breeding pairs among these birds to ensure long-term self-sustainability. Given the delayed breeding strategy inherent in all crane species, this goal will take more time to realize. Nevertheless, early results favor success in this regard, as well. During the summer of 2006, about ten pairs exhibited consistent association with one another and their defended territories; five of these couples incubated eggs. Unfortunately all these early clutches were lost, an event, albeit heartbreaking, typical for young, inexperienced breeding cranes. Patience offers just rewards, however, if we remember that the First Family parents were among pairs who lost eggs in earlier nesting attempts in 2005 and April 2006. They did, indeed, transform their mistakes into competence. The next few years will determine if their ultracrane classmates will follow suit.

On October 5, the 2006 Operation Migration team left Necedah NWR with four ultralight aircraft leading a flock of young cranes to Florida.

Five more are scheduled for reintroduction through Direct Autumn Release. Although all 23 birds are special in their own right, one single-minded female, Number 602, has added significance: her parents were also ultracranes. She is the surviving chick from a Necedah clutch that was left unattended by the naïve pair in April. A strong-willed and enthusiastic flyer, Number 602 also embodies an important link in the chain of baby-step successes that epitomize the eastern migratory flock reintroduction.

That autumn, however, many hopes and dreams rested with a more diminutive flock: the First Family. Carefully guarded as they were by WCEP "grandparents" throughout the summer, humans and cranes alike were forced to incrementally deal out independence as the colts approached fledging. This is, perhaps, a parent's time of greatest joy and apprehension; for the First Family, this time came in mid-September. Once the chicks could gain a little altitude, the family took advantage of increased mobility by foraging a bit farther afield. Although the young birds were not expert fliers by any measure at this time, their innate urge to explore their world would sometimes outstrip Mother Nature's good sense, and occasionally they would be separated from their parents by a quarter-mile or more. Even so, routine weekly tracking surveys of September 3 to 9 recorded both chicks feeding contentedly with their parents. Days later, the smallest chick of the two had disappeared. A frantic air and ground search yielded no results, and after many weeks of desperately scanning the horizon for the little bird, everyone at Necedah was forced to accept the truth. She was gone. Years of monitoring the Wood Buffalo-Aransas population has taught us that the survival of twins in the wild is a rare event. However, this knowledge was about as comforting as the hollow expression "such is life." The recovery of the whooping crane has been plagued with roller-coaster peaks and valleys from the beginning, and the only unfailing guarantee is that this loss will not be the last disappointment.

As we strive toward the golden ring, it is important to remember the larger picture — one thousand whooping cranes is just not enough to weather near eternity. Many ornithologists perceive sandhill cranes as merely "comfortably abundant," but their numbers approach half a million. Even so, this plentiful species was previously extirpated from many US states by unregulated hunting and habitat loss. Like the whooping crane, it survived primarily because remote northern breeding sites remained beyond the reach of most human intrusion. Despite present population numbers, sandhill cranes are also not immune to future anthropogenic influences. Eighty percent predictably use specific stopover habitat on a small stretch of the Platte River, where habitat alteration or communicable disease could destroy a measurable portion of the global population without warning.

The whooping crane is, indeed, more fragile. When sandhill cranes rebounded from decimation a century ago, whooping cranes did not. They

were certainly far less numerous at the time, but attributes of their life history, such as greater reliance on wetlands habitats and potentially lower lifetime reproductive success, may also have conferred substantially more susceptibility to environmental change and other human whims. Although we have amassed considerable experience with whooping crane biology in 60 years of observation, study, hand rearing and migration, we have not changed the bird itself. The risk factors that they faced in the past, they will still suffer in the future. Whooping cranes also face uncertain genetic prospects due to the bottleneck that we forced them to traverse. More recently, our 21st-century reality has contributed additional hazards. The global climate is changing and predicted effects on northern wetlands and coastal salt marshes do not bode well for the species. Furthermore, the remote, once-pristine northern breeding habitat that sustained them through disaster is no longer out of our reach. On the contrary, the riches that these lands possess below the muskeg are all too tempting in this political environment that favors exploitation at considerable cost to the natural world. Undeniably, preserving the whooping crane will require more stewardship than merely reintroducing the requisite number of individuals and expecting them to fend for themselves.

The greatest threat to most crane species, including brolgas, is the loss and degradation of wetland habitats on which they depend.

Then, why should we do it when it would be easier to let it go? What would compel us to invest considerable time and money to save this single bird? After all, the 2006 World Conservation Union Red List of Threatened Species presents 16,118 other plant and animal species that could be deemed worthy of the same consideration. Global economics will ultimately prevent us from rescuing everything. Why, then, save whooping cranes?

The roots of conservation biology have grown from the human desire to satisfy our own needs and preferences. Game species management still serves to maintain an adequately robust population in order that the required number can be killed without tempting extinction. Some species are managed — those that provide good sport — and some are not. Even among ecologists, economists and political analysts, the most immediate justifiable reasons for saving a species are often instrumental or utilitarian. As such, plants and animals have measurable value, and the right action to follow is whatever returns the greatest interest upon investment.

Ecologists would have no reticence when recommending the whooping crane's salvation. Most crane species occupy the upper trophic levels in precious wetland ecosystems. In other words, their position as top predator — prey being amphibians, fish and various invertebrates — confers considerable stability to the entire community. In addition to wetlands being home to a diverse array of other life forms, they double as purveyors of water quality benefits: they are effective assimilators of waste and they function to regulate water flow in rivers and streams. Crane wetlands are of added importance because the birds require large blocks of habitat. Whatever else occupies these extensive marshes will be saved vicariously under the cranes' broad wings. And because these wetlands are distributed both in the north and south and along a migratory corridor that connects summer and winter habitats, conserving the whooping crane and its critical needs fosters international cooperation. Certainly no single species recovery program has demonstrated such success in this regard. What began with desperate aerial searches to find the northern breeding grounds has evolved into the vast network of dedicated workers represented by the 10 members of the International Whooping Crane Recovery Team.

One could also demonstrate justifiable cause to save the whooping crane on solid economic grounds. Recreation and tourism have become a burgeoning element of the North American economy, and years of economic growth have provided many North Americans with disposable income. People must find somewhere to spend their leisure time and money, and they often choose to spend it on whooping cranes. Each year, 150,000 people visit Necedah NWR and 30,000 tour the nearby International Crane Foundation. In winter, 80,000 people escape the cold at Aransas NWR, 17,000 of whom spend up to $35 each to catch a glimpse of cranes from their Intracoastal Waterway tour boat. Annual tourist revenues attributed to the cranes' presence represent $6 million to Rockport, Texas, and $15 million to communities along the Platte River.

In 1984, University of Wisconsin researchers used contingent valuation — which quantifies nonmarket value of a natural resource based on people's willingness to maintain it and their perception of the overall cost of its hypothetical loss — to estimate an existence value for whooping cranes. Their results demonstrated that the whooping crane had an annual worth of about one billion dollars to the people of the United States. Considering projected budget expenditures of about six million dollars per year to fund the recovery program, saving whooping cranes seems a reasonably good bargain.

For certain, local economies benefit considerably from the presence of an endangered species in their midst, particularly one so marvelous as the whooping crane. Moreover, this value is clearly amplified through public awareness and a mounting, transcendent appreciation for the species. It is, perhaps, because of these intangible aesthetic qualities that the whooping crane is now identified as an international symbol of conservation. What is more, it is not one that is failing but one that exemplifies success. Let us appreciate those few insightful humans who recognized the true worth of the whooping crane before it was lost to those of us who came after them. Even then, they knew the

Every whooping crane led from Wisconsin to Florida brings the species one step closer to recovery.

considerable cost could they no longer admire the whooping crane's beauty or hear its triumphant bugle as it returns home in spring.

Maybe it is our appreciation for this intangible value, one that cannot be quantified through legislation, trophic levels or tourist dollars, that has changed measurably in the last century. Empirically, the First Family's destroyed chick is no more valuable than the Saskatchewan baby that was killed without remorse in 1922; in truth, whooping cranes may be more abundant now than they were then. But somehow today in the collective consciousness of humankind, the Wisconsin chick possesses greater worth. Her continued existence was important, and her death was deeply mourned. Perhaps this speaks volumes for some degree of enlightenment among the technological and economical aspects of

After an absence of almost a century, whooping cranes once again grace the skies of the eastern migratory flyway.

© www.operationmigration.org

human cultural evolution since whooping cranes hit rock bottom. Considering the pervasive mass extinctions that our small blue planet is presently suffering at the hand of humans, perchance this newfound clarity will provide some hope for our future.

Author's Note

On November 19, 2006, the First Family — two ultracranes and their one re-maining wild-born chick — left Necedah National Wildlife Refuge with the wind at their backs in an enormous wave of migrating cranes. Twenty-one days later, the three birds touched down on familial wintering territory in Pasco County, Florida; the season passed without incident.

But such is not always the fortune of whooping cranes, whose return from the brink of extinction is marked with tragedy that is ever only a wingtip away. And so it was on February 1, 2007, when violent thunderstorms swept through the darkness at Chassahowitzka National Wildlife Refuge. Eighteen young cranes were roosting peacefully in the shallows, their 76-day journey behind an ultralight aircraft still lingering in recent memory. Suddenly, they were stunned by a bolt of lightning and overcome by rising water. Seventeen birds perished that night. One young crane escaped, only to die later from the complications of solitude.

Yet this species is distinguished by a tenacity that is rivaled by few. Spring migration began like clockwork in late February. Among the birds returning to summer in Wisconsin were the First Family parents leading their wild-born chick, who had so recently become the sole survivor of the eastern flock's 2006 breeding season. They arrived in Necedah by March 20, and within days the adult pair was surveying their breeding territory in anticipation of their 2007 brood. Their chick was already fraternizing with other young cranes nearby. The circle was complete, and hope renewed.

All cranes communicate their desires and intentions through elaborate visual displays, the most beautiful being their dance.

Chapter 6

Species Profiles

Species Profiles Introduction

Cranes, as a group, share many characteristics. They are long-legged and broad-winged; they vehemently defend their family and resident space; and they celebrate springtime with clear voice and elaborate display. It would be iniquitous, however, to imply that they all do so in a similar manner, for this is not the case. Cranes represent 15 distinctively different species, each of which exhibits its own interpretation of cranedom. Hence, the purpose of the following profiles is to showcase each one apart from the other so that their unique beauty and charm may be thoroughly appreciated.

Appearance

This introductory section simply describes the most notable identification characteristics for each species, including height, weight and wingspan, and plumage and bare skin coloration. Plumage of juveniles and downy chicks is also discussed briefly.

Distribution and Seasonal Movements

The mobility of birds is unparalleled in the animal kingdom. While their seemingly unbounded freedom may pique our imaginations, it makes a brief description of geographic range and seasonal movements somewhat problematic. Hence, these sections provide only the barest outline of breeding and nonbreeding distributions and routes that the birds may take between the seasons. Some species have also been divided taxonomically into two or more subspecies, typically based on differing physical characteristics or behavior and disparate geographic ranges.

Habitat

Cranes are often perceived as waterbirds, but this is not true of all species. This section describes in some detail the habitat types that each species prefers during their annual cycle of nesting, wintering and migration.

Food Habits

All cranes are omnivorous; however, their food preferences do vary somewhat between species. This section outlines the favorite foods of each crane, and provides some description of how they procure their meals.

Behavior and Social Interactions

All animals communicate with each other through an immense number of subtle to almost ostentatious behavioral signals; among cranes the most important are, unquestionably, song and dance. Consequently, this section summarizes frequently used vocal and visual signals, with special emphasis on how the behaviors remain distinctively different between species. Also included are descriptions of specific combinations used in defense of territory, which, from a crane's point of view, is of considerable importance.

Reproduction

This section describes both chronology and behaviors inherent in a typical breeding season from nest construction through incubation, hatching, chick rearing and fledging. Details characteristic of particular species are highlighted.

Conservation Status

Conservation biologists use many criteria to evaluate the health of species populations, and frequently apply terms such as "threatened" to delineate those that require special consideration. The World Conservation Union summarizes these criteria in a rating system that incorporates not only the overall number of surviving individuals, but also their geographic distribution, population trends, number of breeding pairs and the nature of their greatest threats. Their rankings as listed from lowest to highest conservation need are: least concern, near threatened, vulnerable, endangered and critically endangered. A critically endangered species is one that has a 50 percent chance of becoming extinct within 10 years. The Endangered Species Act (ESA) uses similar rank names. This section also provides a more precise estimate of numbers of individuals existing worldwide, and outlines factors contributing to observed population trends. Finally, conservation initiatives are discussed with an emphasis on future efforts and potential for recovery.

Grey Crowned Crane
(Balearica regulorum)

Crowned cranes are considered to be the most primitive of crane species. Unlike the others, they have stout, rather than spearlike, bills and their head is topped with a luxuriant crown of stiff golden feathers. Their long hind toes allow them to grasp branches as they roost in trees. Crowned cranes also lack the long, coiled trachea (windpipe) that produces the characteristic bugling calls of other cranes. The grey crowned crane is the national bird of Uganda.

Appearance
Height: 40–45 in (100–110 cm)
Weight: 6.6–8.8 lbs (3.0–4.0 kg)
Wingspan: 70–80 in (180–200 cm)
Adult primarily gray with prominent white cheek patches, short grayish bill and red wattle-like gular sac (throat pouch) that can be inflated when vocalizing. Golden crown tipped with black. Wings appear broadly white with darker flight feathers when open but form large pale triangular patch when held closed. Males and females indistinguishable, except males are slightly larger in size. Juvenile also predominantly gray with feathers broadly edged with rufus or buff. Neck and head rufus with pale cheeks and short, brownish crown. Chicks abundantly covered with soft golden down.

Distribution
Grey crowned cranes occur in east-central and southern Africa. Two subspecies are recognized. East African crowned cranes (*Balearica regulorum gibbericeps*) range from eastern Zaire, northern Uganda and Kenya south to Zimbabwe, Botswana and Namibia. South African crowned cranes (*B. r. regulorum*) breed in Zimbabwe and South Africa; however, nonbreeding birds have been observed in southern Angola. Grey crowned crane distribution has changed little historically except, perhaps, for some localized range reductions in the 1980s and 1990s.

Seasonal Movements
Although grey crowned cranes do not migrate, they do make short seasonal movements in search of food, appropriate nesting habitat and open water. This behavior is more prevalent during the dry (nonbreeding) season — and among populations that typically reside in the drier eastern part of their range — when local water sources become less dependable.

Habitat
Historically, grey crowned cranes used wetlands for breeding and adjacent open grassland and savanna habitat for feeding. However, with much of the traditional habitat converted to agriculture, this species now frequents human-modified areas such as pasture and croplands, fallow fields and irrigation sloughs. Nonetheless, they still require marshlands in which to raise their young.

Food Habits
Grey crowned cranes are omnivorous. They prefer fresh, succulent tips of grasses; seeds; insects, such as grasshoppers and cutworms; and small vertebrates including frogs, lizards and crabs. They also forage among grazing livestock or in croplands for soybeans and millet, where they generally show little concern for human activities.

Behavior and Social Interactions
Grey crowned cranes produce richly harmonic, mournful, low calls. The most frequently heard vocalization is a plaintive *ya-oou-goo-LUNG*, accented on the last syllable, produced as air is forced out of the gular sac. Pairs also give booming unison calls that may last more than a minute. During these displays, participants fall in and out of synchrony with each other as each bird contributes calls at its own rate. Mated pairs also give low, snoring *purrrr* calls to each other and their chicks.

Bowing, jumping, stick tossing and wing flapping typify the dance of grey crowned cranes. Dancing usually begins with considerable simultaneous head bobbing, which is then followed by side-by-side high leaps in the air, often 6 to 8 feet (2 to 2.5 m) off the ground. Dancing is most frequently associated with courtship, but can occur spontaneously as if to relieve tension and aggression.

Grey crowned cranes vigorously defend their nesting territory against intruders. They are particularly diligent against other cranes and geese; however, they are notably fearful of wattled cranes and tend to treat them with caution. The display begins near the territory boundary, where defending birds slowly advance in unison on the intruder with their wings spread wide while uttering several loud disyllabic *ka-wonk* calls. Invading birds usually withdraw quickly.

Reproduction
Breeding typically begins at about three years of age. It is most often associated with the rainy season (October to May), except in eastern Africa, where nesting can occur year-round. Both parents participate in constructing their large, flattened nesting platform of uprooted grasses and sedges, which is well concealed by wetland vegetation. Females lay two to four light blue eggs that hatch in about 30 days. Females and males share incubation duties; at night, the nonincubating bird roosts in a nearby tree. Young cranes are active and robust within 12 hours of hatching and are able to follow their parents as they feed. Unlike other cranes, crowned crane chicks

tend to feed in heavy vegetation cover. As the sun sets, parents carefully lead their chicks back to the nest area, where they huddle together for warmth. Chicks fledge at about 50 to 60 days of age but usually remain in their parents' care for about 10 months, after which they join other juvenile cranes in small nomadic feeding flocks. Typically, only one chick from each clutch of eggs survives beyond three months of age.

Conservation Status
East African crowned crane: population declining.
South African crowned crane: population stable.
World Conservation Union Rating: least concern.

Grey crowned cranes are the most abundant African crane — numbering about 90,000 birds; nonetheless, they are declining steadily, with losses being an estimated 10 percent since the 1980s. The reasons for population declines are pervasive, but generally involve loss and degradation of preferred wetland breeding habitats and conversion of grassland and savanna for agriculture. Other problems include heavy pesticide use and alteration of hydrological patterns for irrigation. Hunting, live trapping for trade and egg collecting are also of concern, although this species' revered status in some areas, particularly Kenya, Namibia and Zambia, has provided a degree of protection.

Conservation programs designed to avert population declines are most

well established in Kenya, where community-based wetland protection initiatives are encouraging sustainable resource and land use. The grey crowned crane has been a critical component in these programs because its cultural significance as a sacred bird has both demonstrated need and provided motivation for community action. Although not globally threatened at present, continued rapid human population growth and degradation of habitat may ultimately pose serious threats to grey crowned cranes. Fortunately they breed readily in captivity, a distinct benefit should recovery programs be required in future.

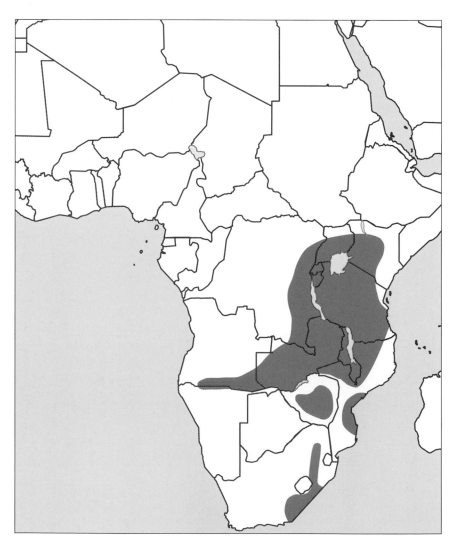

Africa
Year-round distribution

Black Crowned Crane
(Balearica pavonina)

In some West African countries, such as Mali, people traditionally keep black crowned cranes as pets to clear homes and yards of spiders, insects and small reptiles. Unfortunately, generations of illegal trade in live birds has driven some local wild populations to near extinction.

Appearance
Height: 40–42 in (100–105 cm)
Weight: 6.6–8.8 lbs (3.0–4.0 kg)
Wingspan: 70–80 in (180–200 cm)
Adult primarily black with prominent red and white cheek patches and small, black wattle-like gular sac that is inflated during vocal display. Head crowned with large tufts of stiff golden feathers. Elaborate black, fringe-like feathers extend down neck. Distinctive white wing coverts appear as large triangular patches on folded wing and broad bands at leading edge of wing in flight. Long hind toe (hallux) can be used to perch in shrubs and trees. Sexes identical in plumage; however, male is somewhat larger. Juvenile blackish with feathers edged with rufus on upper body and buff below. Head and neck rufus with yellowish cheeks; buffy crown, short and spiky. Chicks covered with soft golden down.

Distribution
Black crowned cranes currently inhabit the Sahel and Sudan savanna regions of Africa; however, historically, they were far more widespread, being found in at least 27 countries from the Atlantic coast through west and central Africa. There are two recognized subspecies. Sudan crowned cranes (Balearica pavonina ceciliae), most abundant of the subspecies, are found in eastern Africa in the upper Nile River Basin. West African crowned cranes

(B. p. pavonina) occur in the western part of the range from Senegal to Chad. They are fragmented into eight or more separate populations, the largest being located along the Niger River.

Seasonal Movements
Black crowned cranes are considered nonmigratory by nature, and seasonal movements are generally limited to flights of 30 to 60 miles (50 to 100 km). These flights occur primarily during the rainy season as flocks seek out small temporary wetlands formed through the localized flooding of lowland savannas. Large flocks have been observed in northern Cameroon and near Lake Chad in northeastern Nigeria.

Habitat
During the dry (nonbreeding) season, black crowned cranes flock in large permanent wetlands, such as freshwater marshes and along edges of lakes and rivers. Smaller temporary wetlands are sometimes occupied by breeding pairs during the rainy season. These cranes frequently use grassland habitats or abandoned agricultural fields for feeding, and may roost in acacias or other trees if they are located near water.

Food Habits
Black crowned cranes eat a variety of food types, including grasses, seeds, insects, millipedes, crabs and small reptiles. They usually forage away from water in upland habitats, and often favor agricultural areas near livestock, where they can find insects in abundance. Crowned cranes stamp their feet while foraging, perhaps to disturb concealed insects. They sometimes eat seeds from planted crops but, unlike grey crowned

cranes, are not considered nuisances by local farmers.

Behavior and Social Interactions
The black crowned crane's voice is somewhat plaintive and goose-like, quite different from the bugling calls of most other cranes. Their characteristic call is a booming honk that is produced by pushing air out of an inflated gular sac. It is repeated frequently and can be heard for a very long distance on a calm day. Pairs also give a unison call that comprises a series of monosyllabic honks; females' calls are higher-pitched than males' calls.

Male black crowned cranes initiate the courtship dance. After several minutes, the female joins in and they hop, bow and jump together frantically. They are territorial when nesting, and both males and females chase nonbreeding cranes and other bird species from the nest site. Other breeding cranes are further discouraged by the male's defense display, in which he arches his neck, points his bill groundward and spreads his feathers. Threats may also include bouts of symbolic "false preening" in which males turn their heads backward and perform exaggerated preening motions. Attack is rare, the display generally being sufficient to drive away rivals.

Reproduction

Black crowned cranes begin breeding at about four years of age. Breeding typically occurs from July to October but varies seasonally in response to rain. Breeding cranes build large, circular platforms of grasses and sedges in well-vegetated wetlands, choosing a location where the nest is surrounded by shallow water to provide some protection. Nesting territories vary considerably in size, from about 200 to 900 acres (80 to 360 hectares), depending on local densities of breeding cranes. Females typically lay two to four light blue or pinkish eggs. Incubation lasts 28 to 31 days, with both parents taking turns incubating and guarding the nest. When the female leaves the nest to feed, the defending male will sound a loud alarm call if he perceives a threat. Chicks remain near

the nest for a few days, then follow their parents to the drier parts of their territory, where they learn to feed for themselves. Their first flight occurs at about age 60 to 100 days, although they remain with their parents for many months afterward.

Conservation Status

Sudan crowned crane: population declining.
West African crowned crane: population declining.
World Conservation Union Rating: near threatened.

Black crowned crane populations are declining throughout their range; however, their reduction in numbers is most dramatic in their western distribution. Recent estimates suggest that West African crowned cranes may number less than 13,000 birds, with some fragmented populations already locally extirpated. In Nigeria alone, this subspecies — considered the national bird — comprised more than 15,000 birds in the early 1970s, but is represented today by a few scattered individuals. Sudan crowned cranes probably number 30,000 to 50,000 individuals.

The primary threat to black crowned cranes is the loss and degradation of preferred wetland and flooded grassland habitats. Causal factors are not simple, being a combination of many

elements including extended drought in the Sahal and sub-Sahal regions; escalating human population; transformation of habitat for agriculture; and modification of river drainage basins for dams and irrigation. Compounding these factors are declines due to widespread unlawful hunting and capturing of black crowned cranes for food and the black market pet trade.

Although the species remains in jeopardy, recent conservation efforts have fostered some interest in declining black crowned crane populations. The 1992 international crane conference in Nigeria spawned the establishment of the Black Crowned Crane Working Group, mandated to review strategies to protect the species. In addition, several national parks and wildlife management areas containing crane habitat have been established, and fieldwork to survey population numbers is ongoing. Despite these changes, the future of black crowned cranes remains uncertain. Cranes often use habitats outside protected areas, particularly during breeding season, and law enforcement is generally ineffective under these circumstances. Furthermore, conservation efforts have been hampered by a lack of knowledge of this species' ecology and accurate estimates of population size and distribution.

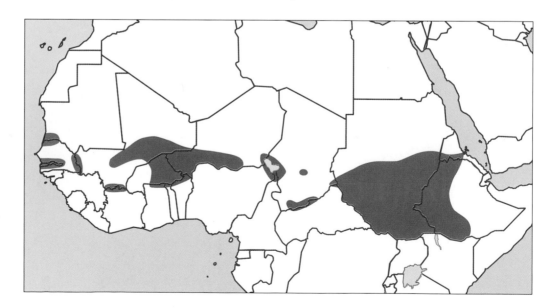

Africa

■ *Year-round distribution*

Demoiselle Crane
(*Anthropoides virgo*)

The delicate demoiselle crane is the smallest crane species. These lovely birds were first brought from the Russian steppes to the royal court of France during the reign of King Louis XVI and Queen Marie Antoinette. The queen was so enchanted by their appearance that she named them *demoiselle*, meaning maiden, or young lady, in French. The dance of the demoiselle crane has been described as an avian ballet.

Appearance
Height: 33–37 in (85–95 cm)
Weight: 4.4–6.6 lbs (2.0–3.0 kg)
Wingspan: 60–67 in (150–170 cm)
Adult pale bluish gray with black head and neck and greenish yellow bill. Prominent narrow white patch behind red eyes extends into long feathered tufts at nape. Long, dark feathers on lower foreneck fall over breast. Flight feathers of wings black. Males slightly larger than females. Juvenile duller gray with pale brownish head and neck. Ear tufts less prominent; eyes brown. Downy hatchlings primarily buff with somewhat darker wings and paler underparts.

Distribution
Demoiselle cranes nest primarily in central Eurasia from the Black Sea east to Mongolia and northeastern China. Small, isolated breeding populations are also found in Turkey and in Morocco's Atlas Mountains. Historically, their breeding range was much more extensive, particularly in northwestern Africa (Tunisia and Algeria) and Europe (Romania and, perhaps, Spain). No subspecies have been identified.

Seasonal Movements
Most populations of demoiselle cranes are migratory except, perhaps, the fragment population in Turkey. Birds breeding in western Eurasia journey south to winter in eastern sub-Saharan Africa from Sudan west through Chad; breeding populations from central Asia east to northeastern China migrate south to wintering grounds in the Indian subcontinent and, occasionally, Bangladesh and Burma.

Habitat
Demoiselle cranes prefer dry grassland habitats such as savannas, steppes and semiarid deserts, often close to streams, small lakes or wetlands. They may nest as high as 10,000 feet (3,000 m) in mountain valleys. Recently, demoiselle cranes have become adapted to using croplands for nesting, particularly where preferred steppe habitat is being converted for agricultural use. Breeding in human-inhabited areas is risky, however, because they are frequently shot or poisoned if considered pests by farmers. In winter, demoiselle cranes seek out wet grassland and savanna habitats, winter croplands or riparian wetlands near acacias or other trees.

Food Habits
Preferred foods include seeds of grasses and other plant materials; insects, particularly beetles in summer; and small vertebrates such as lizards. During migration and on the wintering grounds, large flocks frequently forage in cultivated fields, where they may do considerable damage to cereal grain and legume crops.

Behavior and Social Interactions
Demoiselle crane vocalizations typically comprise short syllables and somewhat guttural notes. The typical *garrooo* call is only three to four seconds in duration; the call's loudness, however, ranges from 78 to 82 decibels. Evidence suggests that individuals with louder calls may be able to maintain larger territories. The unison call is somewhat low-pitched and grating, and consists primarily of a series of single notes. The female usually extends her head backward as she begins calling; the male points his bill skyward as he responds.

Dancing demoiselle cranes are most often observed in spring just after the birds arrive on the breeding grounds, but the behavior can occur through autumn. When dancing, these cranes bow, and jump up and down with half-open wings and outstretched neck while uttering a guttural call. They frequently toss small objects into the air. Occasionally, they can be seen dancing in large numbers, with other birds forming a loose circle around the dancers. Dancing birds flourish their ear tufts and neck.

plumes and fan their tails. Sometimes, dancers exchange places with "spectators" and the display continues until the entire group takes flight, circles and alights on the ground to form small groups or pairs.

Demoiselle cranes defend their nest deliberately, and will call loudly as they attack intruders by striking at them with their bill, wings and claws. Leaping while slashing with claws has been shown to be an effective attack against aerial predators, including falcons (*Falco* spp.). Also, a captive demoiselle crane reportedly killed a man by stabbing him through the eye. Despite their small size, they appear equally fearless of large predators such as dogs, foxes, eagles (*Aquila* spp.) and bustards (*Otis tarda*).

Reproduction

Demoiselle cranes may breed at two or three years of age, perhaps the earliest of all crane species. They nest primarily in April and May. Breeding cranes select their nest site in patchy, somewhat open, habitat, usually near water where the vegetation is high enough to conceal their nest but short enough for the incubating bird to survey its surroundings. Breeding densities are low; nests are typically 1.0 to 2.5 miles (1.5 to 4 km) apart. The nest shows little preparation, simply being a well-camouflaged aggregation of loose pebbles, grasses and roots. Females typically lay two greenish brown eggs that are blotched with purple. Incubation lasts 27 to 29 days. Females perform most of the incubation; males sometimes call or dance nearby to distract intruders or potential predators. They often use an injury-feigning distraction display with their head bowed low and wing tips dragging on the ground. Demoiselle crane chicks are highly mobile; families move away from the nest shortly after chicks hatch, and often travel considerable distances on foot in search of food. Young fledge in mid-summer at about 55 to 65 days of

age but remain close to their parents. Family groups usually assemble into large flocks in preparation for their late-summer migration. Young stay with their parents during migration and presumably through winter.

Conservation Status

Population is stable overall, but some populations declining.
World Conservation Union Rating: none.

The demoiselle crane is the second most numerous crane, with an estimated world population of 200,000 to 240,000 individuals; unfortunately, the species is not beyond threat. Although it remains abundant throughout most of its historic range, the demoiselle crane was extirpated from the Iberian peninsula, Balkan peninsula and parts of North Africa and western Eurasia in the past century. Populations in Turkey and the Atlas Plateau of Morocco may now total as few as 100 birds. Declines have also been noted in the eastern Eurasian and central Asian populations, which have traditionally been considered strongholds.

Principal threats are loss and degradation of preferred habitat on

both nesting and wintering grounds; of particular concern is the Eurasian steppe, where grasslands are recently giving way to agricultural development. Conversion of prime breeding habitat to agriculture has compounded effects through loss of potential nest sites and illegal shooting, intentional poisoning and death from pesticide poisoning, which ensues as some breeding birds feed extensively in croplands. Furthermore, about 5,000 demoiselle cranes (10 to 15 percent of the total migrating population) are killed annually by sport hunters as they travel through Afghanistan and Pakistan.

During the past 20 years, the demoiselle crane has benefited from some limited conservation initiatives that provide localized legal restrictions, protection of important habitat and development of survey and monitoring programs of threatened populations and along migration routes. In addition, they breed well in captivity, and programs mandated to reintroduce the species into areas where it has been extirpated, or where it occurs in critically low numbers, have been considered.

▢ Breeding distribution
▢ Nonbreeding distribution
→ Migratory route

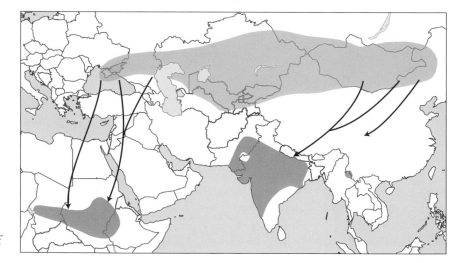

Blue Crane
(Anthropoides paradisea)

Blue cranes are the national bird of South Africa, a country now home to more than 99 percent of the entire population. The species is revered by the Zulu Nation peoples, whose traditions insist that only those of royal blood are allowed to wear the blue crane's beautiful feathers.

Appearance
Height: 45–47 in (110–120 cm)
Weight: 10.8–11.7 lbs (4.9–5.3 kg)
Wingspan: 70–80 in (180–200 cm)
Adult bluish gray with somewhat darker head topped with white. Ashy gray feathers on cheeks, ear regions and nape fluffed during threat displays, conferring cobra-like appearance when angry. Eyes dark brown; bill pinkish. Flight feathers of wing dark gray. Secondary wing feathers very long and drape elegantly toward ground with wing closed. Males and females identical. Juvenile pale gray with yellowish tinge on head and upper neck that fades to gray at age three months. Wing feathers not elongated. Downy young gray with yellowish head and neck and paler underparts.

Distribution
Blue cranes are endemic to (found only in) southern Africa, thus having the most restricted range of all 15 crane species. Most of the population currently resides in South Africa. A very small disjunct population — about 80 individuals — also occurs in northern Namibia. This species was probably never widespread; however, it likely occupied other regions in southern Africa, including Lesotho and Swaziland, where it is now extremely rare. There are no subspecies described.

Seasonal Movements
Although they are nonmigratory, blue cranes make localized seasonal movements within South Africa across elevation gradients. They typically breed in higher-elevation grasslands, and move to lower-lying areas during the nonbreeding season. In addition, vagrant individuals have been observed in Botswana and Zimbabwe, outside their usual distribution, suggesting some seasonal wandering.

Habitat
Blue cranes are most frequently associated with dry grasslands and other upland habitats, but they will also use wetland habitats for feeding and roosting. They often nest at elevations of 4,200 to 6,000 feet (1,300 to 1,800 m), where large predators are less common and disturbance by humans or livestock is somewhat less. Nonetheless, blue cranes often forage in agricultural areas, particularly on grazing land used by small livestock and in fields cultivated with cereal crops.

Food Habits
Blue cranes eat a diverse variety of food types. Plant materials include seeds of grasses and sedges, roots and tubers. They also eat insects (particularly locusts and grasshoppers), worms, crabs, fish, frogs and small reptiles and mammals. In agricultural areas, they regularly feed on waste cereal grain, and may enter feedlots to forage on food made available to livestock.

Behavior and Social Interactions
Blue cranes produce a loud, shrill *kar-rooo* call, given as an alarm by a single bird or in unison by a mated pair. It is often heard at the nest as parents change position for incubation duties. When the male gives his low-pitched, grating unison call, he lifts his wings to expose his dark flight feathers and

arches his neck over his back, then he calls with his bill pointing skyward. Females typically hold their wings at their sides when they participate in unison calling.

Courtship dances, which may last up to four hours, begin with much running in circles interrupted with bouts of loud calling. As the display progresses, both birds tear at grass near their feet and toss it into the air, sometimes kicking or snapping at the bits of vegetation as they fall. The running and calling resumes, followed again by more grass tossing. Mating may or may not occur at this time.

This species is quite aggressive by nature, particularly when nesting, and will generally attack any approaching intruder. The assailing bird spreads its wings, holds its body erect and directs its sharp bill toward the potential threat. If the intruder does not withdraw, the crane lunges repeatedly while stabbing with its bill and kicking violently. It has been noted that waving a stick as defense against an attacking blue crane only seems to increase its ferocity.

Reproduction

Blue cranes first breed at three to five years of age. Breeding season typically occurs from October to December but may extend to March in some years. Pairs usually select a secluded nest site in grasslands with relatively short, thick vegetation. The female lays two heavily spotted grayish brown eggs. Eggs are often placed directly on the ground; sometimes, a nest platform is fashioned loosely out of pebbles or grass. Incubation lasts 30 to 33 days. Although both parents incubate, the male is primarily responsible for defending the nest from intruders, which he does aggressively and with great vigilance. Young cranes remain at the nest for about 12 hours after hatching, and then follow their parents into shortgrass upland areas to feed. When accompanying her young, a female blue crane will be equally as aggressive as her male partner. She feeds the chicks for about 15 days, during which time they learn to pick up food without her aid. Although young blue cranes fledge at about 85 days of age, they stay with their parents until breeding season of the following year.

Conservation Status

Population is declining.
World Conservation Union Rating: vulnerable.

Until recently, blue cranes were not considered threatened with extinction because they existed throughout most of their historical range. However, population surveys revealed that they had suffered substantial declines since the 1980s, as much as 90 percent in some areas. The total population is estimated at 21,000 individuals, although reliable census data throughout their distribution is typically unavailable.

The principal threat to blue crane survival is poisoning. Although illegal in South Africa, intentional poisoning has been an increasingly common assault on cranes feeding in cropfields. Unintentional poisoning also occurs regularly as cranes feed on poisons aimed at killing other species that cause crop damage, or when pesticides are applied directly to croplands. Poisoning occurs most often in late winter or early spring, when cranes feed in large flocks, making them more vulnerable to mass mortality. Although legislation is in place to protect blue cranes from illegal trapping, hunting and poisoning, it is rarely enforceable. Furthermore, fewer than 200 pairs currently breed in nature reserves designed to offer some safety. Fortunately, land management practices that favor the species without detriment to farmers' welfare are well understood. Hence, the preservation of blue cranes will ultimately rely on both the efforts of conservation workers to communicate effectively and the ability of private landowners to consider the ecological and cultural importance of this species.

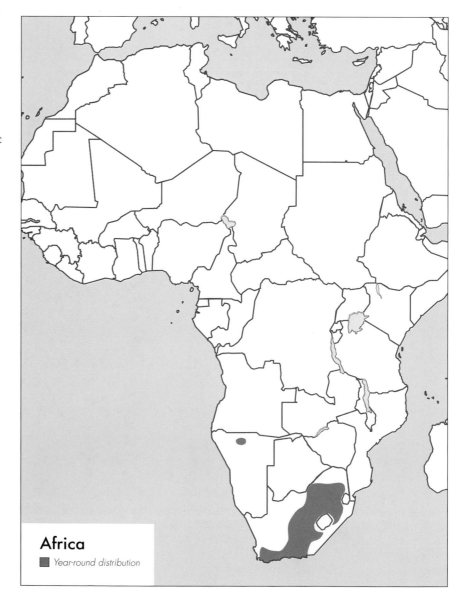

Africa
■ *Year-round distribution*

Wattled Crane
(*Bugeranus carunculatus*)

A recently discovered Mozambique wattled crane population — about 2,500 individuals — represented a substantial addition to the species' overall numbers. Unfortunately, upstream dams have so altered wetlands on the Zambezi delta that wattled cranes now nest only on the fringe of their traditional habitat, where runoff from adjacent uplands still supports patches of spike rush (*Eleocharis acutangula*), a necessary food source for their young.

Appearance
Height: 77–80 in (170–180 cm)
Weight: 18.3–18.7 lbs (8.3–8.5 kg)
Wingspan: 90–102 in (230–260 cm)
Very tall. Adult ashy gray with blackish underparts. Head white with dark gray cap and bright red bare skin from eyes forward to bill. Prominent white wattles hang down from throat. Bill brown and long. Elongated dark flight feathers of wing fall nearly to ground when folded. Male slightly larger than female. Juvenile similar but somewhat tawnier. No bare skin on face; wattles much less developed. Downy chicks buff with somewhat darker, brownish backs and wings.

Distribution
Wattled cranes occur in Ethiopia and in south-central Africa from Zaire, Zambia, and Tanzania south to Botswana and Mozambique. There also are outlying populations in Namibia, Angola and South Africa. Historically, they were more widely distributed across southern Africa; however, numerous local extirpations have created the patchy distribution observable today. This is particularly evident in South Africa, where these cranes were formerly found throughout the country. No subspecies are known.

Seasonal Movements
Most wattled cranes are probably nonmigratory; however, their dependence on wetland habitats forces them to search nomadically for open water on occasion. Populations with access to reliable, permanent wetlands are less mobile than those that use seasonally flooded habitats. Some biologists have considered the seasonal movements of Ethiopian wattled cranes to be truly migratory because of their annual regularity; however, more study is needed to fully understand this.

Habitat
Wattled cranes are highly reliant on wetland habitats for feeding and nesting. Extensive wetland areas in the riparian floodplains of southern Africa's largest rivers, such as the Zambezi and Okavango, are most desired, as are permanent highland marshes. Smaller ephemeral wetlands are also used opportunistically, particularly by young nonbreeding birds.

Food Habits
Wattled cranes feed primarily on aquatic vegetation, including tubers and roots of submerged sedges and water lilies, which they excavate vigorously with their bills. They also eat beetles and grasshoppers and, at times, snails, frogs and other small animals. They occasionally forage in meadows and agricultural fields for grain or grass seeds.

Behavior and Social Interactions
This species may be less vocal than other cranes, but they are capable of producing very loud, shrill screaming calls, the most frequently heard being a single, bell-like *kronk*. This note is sometimes repeated (*kronk-kronk-kronk*) and may be given in unison during courtship or when incubating

parents are changing places at the nest. The female typically initiates the unison call with a quick lowering and raising of her head; the male performs likewise. The display lasts approximately three to seven seconds and terminates with the male uttering a series of short notes followed by one long note as he elevates his wings. When dancing, wattled cranes hold their wings high, open their bills and jump well off the ground. They often whirl in circles as they leap.

Wattled cranes are highly territorial when breeding, and will aggressively defend about 250 acres (100 hectares). They maintain these territories throughout the breeding period for both nesting and feeding, long after their chick has fledged. Nonbreeding birds, particularly those too young to breed, often congregate in flocks that may number almost 100 individuals.

Reproduction
Wattled cranes begin breeding at about three or four years of age. Their breeding season is highly variable, typically being dependent on water levels. They prefer to build their nest during flood peaks (usually August and September), when there is less risk of it being overcome by rising water. Nests are constructed of mounded vegetation in an open grass and sedge marshland where standing water is about 3 feet (1 m) deep, creating a wide moat. Wattled cranes usually produce only one egg per year, which is buff in color with brown spots. Both parents incubate for a period of 33 to 36 days, with frequent changes in incubation duty throughout the day. The nonincubating bird often flies some distance from the nest to feed; however, once the chick hatches, both parents are very attentive. The family remains in the nest vicinity for several weeks, and may return to the nest to sleep. Once old enough, the chick is led to nearby grassy meadows to feed; it will remain under its parents' care until the following breeding season. Wattled cranes have the longest egg-laying-to-fledging period (up to 180 days) of any crane species, making the chicks particularly vulnerable to predation.

Conservation Status
Population is declining.
World Conservation Union rating: vulnerable.

Wattled cranes have been declining throughout their range in recent decades; consequently, they are now the rarest of Africa's six crane species. Although the estimated total population of 13,000 to 15,000 individuals appears to have remained constant, this figure was strongly influenced by the 1992 discovery of a previously unknown Mozambique population numbering about 2,500 birds. More

likely, the Zambian core population fell from an estimated 11,000 to 7,000 birds during this time. Only about 250 wattled cranes remain in South Africa; unfortunately, this fragment population is also declining. Less is known of the Ethiopian population, which currently numbers several hundred birds.

The dependence on wetland habitat for nesting and feeding makes wattled cranes highly susceptible to habitat loss and degradation. Principal threats include alteration of floodplain habitat along major rivers — particularly changes in water levels and vegetation types — from hydroelectric power projects and other water development programs. Loss and degradation of smaller wetlands have been attributed to intensification of agricultural practices, damming for water storage, industrialization and urbanization. Widespread irrigation in some areas has caused groundwater levels to drop so significantly that wetlands are no longer useful to cranes. Other threats include

intentional burning of grasslands that kills flightless chicks and mass aerial spraying associated with the tsetse fly control program, which can cause widespread poisoning.

Protected reserves have been established in several key wetland habitats, especially in Zambia, Namibia and Botswana. In South Africa, community-based habitat conservation programs have encouraged private landowners to safeguard the species by managing their lands appropriately; however, no incentive programs exist to promote participation. Studies have demonstrated that availability of suitable nest sites is indeed a limiting factor for wattled crane populations; breeding pairs have been observed using artificially constructed impoundments for nesting where other acceptable nest sites are lacking. This species breeds less readily in captivity than other crane species; hence, it is imperative that conservation and restoration of preferred habitat be given priority.

Africa
■ Year-round distribution

Siberian Crane
(*Grus leucogeranus*)

Wild cranes typically live 20 to 30 years. However, the longest-lived crane on record was a Siberian crane named Wolf, which lived to age 83 at the International Crane Foundation in Baraboo, Wisconsin. Wolf fathered chicks at age 78, which attests to the life-history strategy exhibited by all cranes, which insists that populations persist only if adults are able to raise successfully one or two chicks per year every year of their long lives.

Appearance
Height: 53–57 in (135–145 cm)
Weight: 10.8–18.9 lbs (4.9–8.6 kg)
Wingspan: 83–90 in (210–230 cm)
Adult pure white, except black primary wing feathers. Forehead, face and side of head featherless and red in color. Eyes yellow. Male and female similar in plumage; however, male slightly larger with longer bill. Buffy brown juveniles have rusty heads and necks and somewhat paler throats. Downy chick buff with darker upperparts; blue eyes turn yellow at about six months of age.

Distribution
The critically endangered Siberian crane undoubtedly once had a broader range than at present; there are only two populations remaining. The larger eastern population (probably fewer than 3,000 birds) breeds in northeastern Siberia, but nonbreeding individuals have been observed near the Russia-Mongolia-China border during breeding season. The small western population breeds near the Ob River, east of the Russian Ural Mountains. These breeding grounds were only recently discovered when a pair of cranes was tracked via satellite as they returned north from their wintering site. There once existed a centrally located third population that nested in western Siberia and wintered in India; unfortunately, it was extirpated in the 1990s.

Seasonal Movements
Both populations are migratory. During migration, the eastern population travels approximately 3,200 miles (5,100 km) along river valleys en route to eastern China. About 98 percent of these birds winter at Poyang Lake, China's largest freshwater lake, in northern Jiangxi Province. Siberian cranes from the western population migrate south to a single wintering site on the south coast of the Caspian Sea in Iran. Birds typically arrive on their wintering grounds in October and depart for the north in March.

Habitat
Siberian cranes may be the most aquatic of all cranes, relying almost exclusively on wetland habitat for nesting, feeding and roosting. Breeding habitats include tidal flats, bogs and marshes in open lowland tundra or northern taiga. During migration stopovers, large isolated wetlands are preferred. Wintering cranes inhabit seasonally flooded mudflats in China, but typically use artificial water impoundments and wet rice fields in India and Iran.

Food Habits
On their northern breeding grounds, Siberian cranes consume a variety of plant materials, such as roots, seeds, berries and sprouts of sedges. They also readily eat insects, fishes, rodents and other small animals, particularly early in the breeding season when plant foods are less available. When migrating or wintering, their food choices are somewhat limited, being more restricted to aquatic plants; thus, they feed only opportunistically on aquatic animals.

Behavior and Social Interactions
Vocalizations and behavioral displays of the Siberian crane are rather distinctive. They have a clear, melodious, high-pitched voice that is more whistle than bugle. Calls are often comprised of several dozen repeated syllables: *a-hooya–a-hooya– a-hooya–a-hooya–a-hooya*. Males initiate the unison call with distinctive looping head movements and a continuous series of flute-like notes; the female joins in, bobbing her head up and down with each note. The result is a singsong repeated *doodle-loo, doodle-loo*, with the male contributing the *doodle* and the female the *loo*. Flock members also whisper a pleasing, musical *toya toya* when at roost. Juveniles give a piercing call to encourage their parents to feed them.

Siberian cranes' dances differ somewhat from other cranes'. Although they comprise usual behaviors such as jumping, running and flapping wings, they also include a curious bowing posture in which heads and necks are brought forward from the vertical to be positioned downward and backward between the legs as wings are held high. The male crane typically initiates the dance, which may be most frequently associated with courtship but can occur at any time on wintering or breeding grounds.

Siberian cranes are territorial when nesting, and experienced birds often occupy sites used in previous years. The size of these territories is difficult to ascertain because the birds occur at such low density, due, in part, to their reduced numbers; hence, direct conflict between neighbors is rare. One study reported a breeding density of one pair per 112 square miles (290 square km) in the Yakutia Republic in 1981.

Reproduction
Siberian cranes probably begin breeding at about three years of age. Nesting occurs in May or June. Pairs preferentially choose a site within an expansive wetland with shallow, fresh water and good visibility. The nest is little more than a flat mound of grass and sedges that is elevated about 6 inches (15 cm) above surrounding water. The female lays two eggs that are deep olive in color and spotted with brown and gray. Both males and females incubate the eggs, in turn, for about 29 days. They are very attentive at the nest and both will guard the vicinity; however, the male takes the primary role in defense. If unduly disturbed, the incubating parent will sneak off the nest and perform a wing-drooping distraction display. Chicks fledge in about 70 to 75 days. Only about 10 percent of chicks produced in any given breeding season survive to winter because of high predation on the nesting grounds from dogs and egg-eating birds such as gulls (*Larus* spp.) and jaegers (*Stercorarius* spp.); aggression

between siblings should two eggs in a clutch hatch; and fatalities associated with the arduously long autumn migration. This may be the lowest rate of annual reproductive success in all cranes.

Conservation Status
Population is declining.
World Conservation Union rating: critically endangered.

The Siberian crane is the third-rarest crane species. Siberian cranes were probably never common; however, their numbers and overall geographic distribution have been decimated since the 1800s. The larger eastern population numbers fewer than 3,000 birds. Of grave concern is that this population, representing virtually all of the world's Siberian cranes, winters in a single wetland on Lake Poyang in China. Degradation of this habitat that may ensue with the completion of the Three Gorges Dam on the Yangtze River will undoubtedly be the death knell for the Siberian crane. The western population also poises on the brink of extirpation; recent counts suggest only about nine cranes remain in this group. The central population was extirpated as recently as the 1990s. These birds were known primarily on their wintering grounds, where a 1965 census reported 200 individuals. By 1995, the group included only a mated pair and their chick and another lone adult. No cranes from this population have been reported since 2002.

Although the Siberian crane's breeding grounds are remote, they are currently under threat from oil exploration

and development. Winter logging practices are also of concern. The most urgent threat to its survival, however, is the loss and degradation of crucial wetland habitat required during migration and when wintering. This is most problematic in China's eastern provinces, where increasing human population has caused intensification of agriculture, changes in hydrological patterns by damming rivers, pesticide use, increased industrial pollution and poaching. Also, proposed dam projects on the Amur River, which divides Russia and China, would destroy critical migratory stopover habitat. Sport-hunting pressure is also considerable, and may have been the single most important cause in the central population's recent demise.

Fortunately, conservation measures are taking place on many fronts. International cooperation is increasing; and in 1993, eight range states signed a memorandum of understanding that acknowledged a need for participation in conservation activities. Reserves protecting key habitat have since been established along migration routes and wintering sites, and increased field research is providing critical information regarding this species' life history and population viability. Also, a captive breeding and reintroduction program is underway. With luck and dedication, these efforts may be fruitful. Unless they show considerable success, however, the predicted decline in excess of 80 percent over the next decade may yet become reality.

Asia

■ Breeding distribution
■ Nonbreeding distribution
■ Year-round distribution
→ Migratory route

Sandhill Crane
(Grus canadensis)

Many crane species, including sand-hills, lay two eggs. However, sibling rivalry is so intense that typically only one chick survives to fledge. Parents can sometimes quell the aggression between their offspring by additional feedings, which is why two chicks are raised only in years of abundant food.

Appearance
Height: 35–47 in (90–120 cm)
Weight: 6.6–14.4 lbs (3.3–6.5 kg)
Wingspan: 63–83 in (160–210 cm)
Adult gray with whitish face, throat and cheeks. Crown and forehead covered with red skin. Plumage may appear rusty from iron-rich mud preened into feathers, perhaps to improve camouflage at the nest; coloration disappears in autumn as new feathers replace stained ones during annual molt. Males and females identical, except males slightly larger. Subspecies similar in plumage but vary considerably in size. Juvenile resembles adult, except back and wings edged with cinnamon brown. Downy chick tawny brown with darker upper parts and pale gray throat and belly.

Distribution
Six subspecies are currently recognized. The most populous subspecies are lesser sandhill cranes (*Grus canadensis canadensis*), greater sandhill cranes (*G. c. tabida*) and Canadian sandhill cranes (*G. c. rowani*), which breed through northern North America, from eastern Ontario west to the Pacific Ocean and north to the Arctic, and in eastern Siberia. Three nonmigratory subspecies occur in small isolated populations in southeastern Mississippi (Mississippi sandhill crane, *G. c. pulla*), central Florida (Florida sandhill crane, *G. c. pratensis*) and Cuba (Cuban sandhill crane, *G. c. nesiotes*).

Seasonal Movements
Lesser sandhill cranes migrate from arctic and subarctic breeding grounds to the southwestern United States and north-central Mexico. Greater sandhill cranes, which breed further south at mid-continent latitudes, also winter in the southern United States and northern Mexico. Canadian sandhill cranes migrate south from nesting grounds in subarctic Canada, from British Columbia to northern Ontario, to wintering grounds in coastal Texas, the southwestern United States and north-central Mexico. Some populations of these subspecies winter together. A key spring migration staging area occurs along a 75-mile (120 km) stretch of the Platte River in Nebraska, where up to 80 percent of all sandhills concentrate to rest, renew their pair bonds and restore energy reserves that they require to complete their journey north.

Habitat
During breeding season, sandhill cranes prefer open freshwater wetlands such as muskeg, bogs and sedge meadows in the north, and flooded meadows and marshes at mid-continent latitudes. Migration stopover sites are typically shallow marshlands, wet agricultural fields and river shallows and sandbars. Wintering cranes prefer wetlands associated with coastal areas, small salt lakes and rivers. Mississippi and Florida subspecies often use shallow wetlands and wet pastureland. Sandhill cranes in Cuba are not associated with wetlands, but prefer dry, isolated savanna habitats.

Food Habits
Diet varies considerably depending on location and season. Sandhill cranes will readily consume waste grain such as wheat, sorghum and barley in agricultural fields, particularly during migration and in winter. Tubers, berries and acorns are also commonly eaten. Animal foods include insects, worms, snails and small rodents.

Behavior and Social Interactions
Sandhill cranes produce many variations of a loud rattling or bugling *kar-r-r-o-o-o*. This call may be given in unison by a mated pair to reinforce their pair bond or to drive away potential predators. The female often initiates unison calling with a series of pulsing notes that resemble machine-gun fire. Both birds keep their wings folded at their sides throughout the display. Single-note echoing variations of this rattling call may be produced to warn other cranes of impending danger. Sandhill crane dances are

frequently associated with courtship, although they can occur anytime. They typically include leaping and tossing objects into the air; however, unlike other cranes, sandhills hold their wings completely folded during the display. They also display in defense of nesting territory and to establish social dominance hierarchy, particularly in flocks during prenesting pair formation. Both males and females defend breeding territories that are relatively small — reportedly 8 to 260 acres (3 to 105 hectares) — in size.

Reproduction

Sandhill cranes do not breed until they are about three years of age. They breed in April or May in the north, but somewhat earlier at southern latitudes. Their nest is a large mound of wetland plants placed in standing water where it is surrounded by emergent vegetation. The female lays two buff-colored spotted eggs about two or three days apart. Incubation takes 29 to 32 days. Both parents share duty during the day; however, the female does most of the night incubation. Nests are attended constantly. Parents defend both nest and young, first attempting to drive away intruders by spreading their wings and directing their bill toward the offender, then kicking or stabbing it if it does not withdraw. Nesting success is usually high (70 to 80 percent). If chicks hatch on different days, one parent will lead the older chick away to feed while the second is being tended near the nest. Chicks fledge 50 to 90 days after hatching. Young sandhill cranes remain with their parents until about nine to ten months of age, after which they join nonbreeding flocks until they are old enough to find a mate.

Conservation Status

Most populations are stable; some populations are declining.
World Conservation Union rating: none.

ESA rating: Cuban sandhill crane: endangered.
Mississippi sandhill crane: endangered.

The sandhill crane is not globally threatened. With a total estimated population of approximately 500,000 birds, it is the most abundant crane species. However, two subspecies that occur in Mississippi and Cuba are of grave concern. The Mississippi cranes number only about 100 individuals (25 breeding pairs), most of which were captive-bred birds reintroduced to bolster the rapidly declining population. Also, a few hundred cranes reside in scattered subpopulations in Cuba.

The most important threat to sandhill crane populations is the loss and degradation of critical wetland habitat. Although numerous at present, migratory sandhill cranes are potentially vulnerable to disease because of their propensity to congregate in large numbers at specific times in their annual cycle. In addition to their immense spring staging, about 80 percent of mid-continent populations winter on fewer than 20 salt lakes in west Texas. Large flocks are also prone to exploitation by hunting. In some years, legalized hunting accounts for about 7 percent mortality in migratory populations; for example, 7,700 sport hunters killed 35,706 sandhill cranes in the 2003–04 season. Nonmigratory populations are also threatened by loss of habitat, most frequently through residential and commercial development.

Conservation efforts should focus on retaining preferred wetland habitat, particularly where cranes congregate. Also, low reproductive rates in most cranes entail the need to assess accurately the effects of hunting on migratory populations. Should serious population declines ensue, it may be difficult for this species to recover.

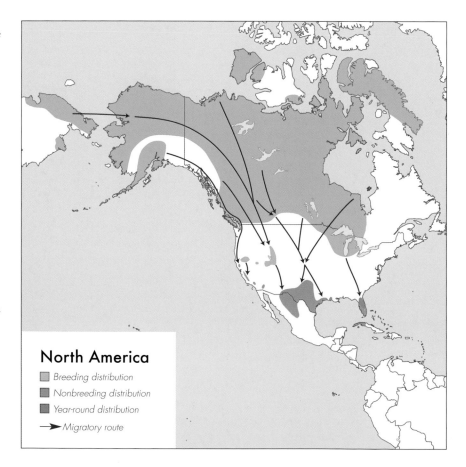

North America
- Breeding distribution
- Nonbreeding distribution
- Year-round distribution
- → Migratory route

Sarus Crane
(Grus antigone)

Male Indian sarus cranes stand up to 6 feet (1.8 m) tall; thus, they are the tallest of all flying birds. This statuesque species is known for its fierce, aggressive defense of family and territory, which includes much posturing and trumpeting followed by directed attack and, sometimes, the destruction of neighbors' nests.

Appearance

Height: 67–71 in (170–180 cm)
Weight: 11.9–26.8 lbs (5.4–12.2 kg)
Wingspan: 78–110 in (200–280 cm)
Adult very tall. Adult predominantly gray. Bare red head has gray cap and small white dot near ear. Upper throat covered with long black bristles. Indian sarus crane has whitish collar. Bill yellowish brown. Male slightly larger than female. Juvenile gray with feathers edged in brown. Head and neck covered in buffy feathers. Downy chick buffy with cinnamon-brown wings and whitish underparts.

Distribution

Three extant subspecies are recognized based on variation in size and plumage. Indian sarus cranes (*Grus antigone antigone*) occur in northern India and western Nepal, having been extirpated from much of their former range across the subcontinent, including Pakistan. One recent record suggests that they may still exist in very low numbers in Bangladesh. The eastern sarus crane (*G. a. sharpii*), which breeds in Southeast Asia, primarily Cambodia and Laos, has also been decimated throughout its historical range; it no longer occurs in China or Thailand. The Australian sarus crane (*G. a. gilli*) is virtually restricted to the Cape York Peninsula in northeastern Australia. Sarus cranes were formerly found in the Philippines (*G. a. luzonica*); however, they are now presumed extinct.

Seasonal Movements

Eastern sarus cranes are migratory, making seasonal flights from their northerly breeding grounds south to the Mekong River delta, where they spend the dry season (December to March). Australian sarus cranes may be partially migratory. Some birds make annual flights of approximately 250 miles (400 km) from breeding grounds on western Cape York Peninsula to winter feeding sites on the eastern side of the peninsula. Indian sarus cranes are nonmigratory; however, they may make limited movements during drought in search of food and water.

Habitat

The habitat requirements of sarus cranes vary seasonally and with geographic distribution. Indian sarus cranes are well adapted to human-modified habitats such as cultivated fields, irrigation ditches and rice paddies, provided that they are not subject to persecution. Otherwise, they use shallow marshes, ponds and other natural wetland habitats. Eastern sarus cranes are more wary of people; however, this species is poorly known and many aspects of their breeding requirements remain obscure. They likely depend on natural shallow wetlands for nesting, but may also frequent sedge meadows and wet grasslands during the dry season. Australian sarus cranes use many habitat types — marshes, upland grasslands and agricultural fields — provided that water is available. Where brolga cranes and sarus cranes occur together in Australia, sarus cranes favor drier habitats.

Food Habits

The diet of the sarus crane varies widely. Consumed plant material includes tubers and bulbs of aquatic plants, seeds, groundnuts, waste grain and rice. They also eat invertebrates such as insects, snails and crustaceans and small vertebrates, including fish and frogs. Sarus cranes have also been observed killing and consuming snakes up to 2 feet (60 cm) in length.

Behavior and Social Interactions

Sarus cranes produce very penetrating whooping vocalizations. Males and females engage in coordinated unison calling, with the more high-pitched female beginning first. When calling, the male holds his neck erect so his bill points somewhat backward and raises his wings to display his white secondary feathers. The female also points her bill skyward but does not raise her wings. The female gives a repeated *tuk-tuk-tuk-tuk-tuk* call, to which the male responds with a trumpeting *garrrrooooa-garrrooa-garrrooooa-garrrooa*. They typically stand side by side when calling, often close enough to touch one another.

Sarus cranes dance throughout breeding season and into winter. Their dance lasts only a few minutes, and is characterized by dashing about and jumping and bowing straight up and down, without the whirling associated with other cranes. Sometimes bits of grass are tossed into the air.

Sarus cranes vigorously defend their territory, which may be 100 to 150 acres (40 to 60 hectares) in size. Rival males face other and "mock-preen" furiously to show their aggressive intentions. In response, the resident bird adopts a defense posture in which he arches his neck, points his bill toward the ground and spreads his wings.

Reproduction

Sarus cranes begin breeding at about three years of age. Breeding periods vary with subspecies, but generally peak nesting occurs during the rainy season. They build a large mound-like nest of wetland vegetation, placed so that surrounding water is about 3 feet (1 meter) deep. The female typically lays two whitish spotted eggs. Males and females share incubation duties, which last 31 to 34 days; however, the female stays on the eggs at night while the male roosts nearby in shallow water. Parents are very attentive to their young, guarding them from predators and helping them to feed. Chicks freeze and remain stationary when alerted to danger by a single parental call and do not come out from hiding until ordered to do so. Chicks fledge at about 90 to 120 days of age but stay with their parents at least 10 months, after which they congregate in loose flocks of same-age birds.

Conservation Status

Population is declining.
World Conservation Union rating: vulnerable.

Sarus cranes have been extirpated from many parts of their former breeding range, including China, Malaysia, Thailand, Vietnam and the Philippines. Their estimated total population is approximately 15,000 individuals; however, they are common now only in parts of northern India, where the species enjoys some protection due to religious significance. Eastern sarus crane populations have been reduced to only 500 to 1,500 birds, with devastating losses in the past 50 years. This species suffers from our lack of knowledge of its life history; in fact, exact location of some breeding grounds has yet to be determined. Australian sarus cranes number fewer than 5,000; however, they are generally considered secure because recent human-induced habitat changes, such as those caused through introduction of cattle, seem to favor the birds. In addition, Australian sarus cranes can outcompete brolga cranes where they coexist.

Principle threats include loss and degradation of wetlands through human population growth, agricultural intensification, industrial development, pollution, pesticide use and warfare. Sport hunting has declined somewhat in recent years — during the early 1900s, sarus cranes were lassoed from horseback in the Philippines — but theft of eggs and chicks for food and the illegal pet trade are still prevalent.

Sarus cranes are legally protected; however, local enforcement of such laws is lacking. In addition, traditions that have played a key role in protecting the species are eroding as increased population puts added pressure on remaining habitat. Seven of ten countries (exceptions being Cambodia, Laos and Myanmar) where the cranes are found have signed the Ramsar Convention treaty, which is mandated to provide a framework for international cooperation on wetland conservation. Establishment of protected habitat areas is underway, and some countries have chosen this species to represent their conservation efforts. Reintroduction plans have been discussed for some portions of their historic range.

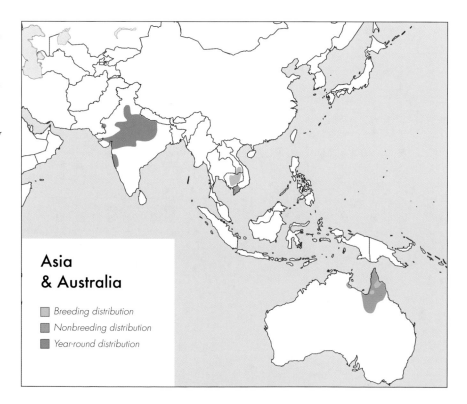

Asia & Australia

- Breeding distribution
- Nonbreeding distribution
- Year-round distribution

Brolga Crane
(Grus rubicunda)

Many tales in Aboriginal Australian folklore describe this crane. One legend recounts Burlaga as a fair maiden forced by her elders to marry an old tribal chieftain. She escaped this fate by fleeing to open country, where she met a young man. As warriors of her tribe descended upon Burlaga and her companion, the two lovers danced as a whirlwind, vanished into dust and reappeared as two large birds, thus eluding their capturers.

Appearance
Height: 59–63 in (150–160 cm)
Weight: 7.9–15.8 lbs (3.6–7.2 kg)
Wingspan: 79–90 in (200–230 cm)
Very tall. Adult light bluish gray with blackish wings. Red unfeathered head has gray cap. Bill gray. Males and females are identical in plumage although males are slightly larger. Juvenile similar except buffy gray head is fully feathered; characteristic bare head acquired at age two or three years. Chick covered with gray down except head and neck, which are tawny colored.

Distribution
Although brolgas were historically divided into two subspecies — northern Australian cranes and southern Australian cranes — ornithologists no longer accept this division. Nonetheless, there are many reasons to consider them as separate populations. Northern birds are widespread in northern and eastern Australia and southern Papua New Guinea; the southern population has a very limited, fragmented distribution in southeastern Australia. Also, there is a sizable region of inappropriate habitat separating them and, as a result, there is little contact between the populations. Historically, southern birds occupied a greater distribution in New South Wales and Victoria; however,

they have declined considerably since the 1890s as human settlement has expanded.

Seasonal Movements
Brolgas are nonmigratory; however, they often make short seasonal flights to find suitable wetland habitat, particularly during periods of extended drought. Also, evidence suggests that brolgas in northeastern Australia may retreat to New Guinea for refuge during droughts. In southern Australia, adults disperse to find supplementary breeding habitat in isolated small, temporary wetlands formed during the rainy season.

Habitat
Many habitat types are used according to season and geographic distribution. During the dry season, northern brolgas prefer sedge marshes, tidal pools and lake edges in coastal areas. They have specialized salt glands located near their eyes that allow them to excrete excess salt taken in as they feed and drink in saline wetlands. They move inland to breed in seasonal wetlands, wet meadows and brackish marshes during the rainy season. Southern brolgas primarily use shallow — less than 20 inches (50 cm) deep — freshwater marshes for breeding. During dry summers, they move to more permanent wetlands, upland pastures and edges of small reservoirs, where they feed in large flocks. Where brolgas and sarus cranes overlap in distribution, brolgas typically use larger, more open wetland habitats.

Food Habits
The brolga diet is highly varied, and they forage in both freshwater and saltwater marshes. Sedge tubers are perhaps the most important food source; however, this species also feeds

opportunistically on insects, crustaceans, mollusks, frogs, small mammals and birds. They may also eat cereal grains and nuts from cultivated fields.

Behavior and Social Interactions
Calls are given frequently while dancing, sometimes by males and females in unison. The male raises his wings above his back and gives a very penetrating and whooping *kawee–kreee–kurr–kurr–kurr–kurr*, to which the female interjects an extended series of *kuk–kuk–kuk–kuk* notes between the male's syllables. Females often initiate bouts of unison calling. Displays are terminated with one or two long, low-pitched calls that are created by air forced out of the gular sac.

Brolgas perform intricate dances, particularly when young, unpaired birds are seeking mates. Males often initiate these displays, which begin with picking up grass or sticks and tossing them in the air. Leaping and spiraling with outstretched wings is then followed with bowing, head bobbing and walking about. Sometimes, many brolgas will dance together. But before group dancing begins, participants usually line up across from each other like square dancers.

Although brolgas typically spend the nonbreeding season in flocks, they are less tolerant of individuals that are not family-group members. If another crane approaches too closely, its neighbor will run toward it with head arched up and forward and wings spread. Even within family groups, males demonstrate aggression toward each other as they compete for higher dominance status.

Reproduction

Brolgas first breed at three to five years of age. In the north, breeding activities begin in November with the onset of rainy season. However, most nesting occurs after January, when water levels have peaked; otherwise, early nests may be drowned as water rises. Southern populations breed from July to November. Nests are large mounds up to 5 feet (1.5 m) in diameter, made from grass and sedge stems placed within dense wetland vegetation. Females usually lay two whitish spotted eggs. Both parents incubate the eggs for 28 to 31 days. Nesting pairs will vigorously attack intruders or potential predators; there are reports of trumpeting brolgas with half-open wings advancing to threaten approaching horses. Young brolgas can run and swim within a few hours of hatching. They fledge at about 100 days of age but stay with their parents for 10 to 12 months. Parents may continue to guard and care for their offspring for an additional year if they do not renest.

Conservation Status

Population is generally stable with local declines.
World Conservation Union rating: least concern.

Currently, brolgas are not globally threatened; however, accurate estimates of total population remain elusive, particularly in northern Australia and New Guinea, where these birds are inadequately studied. Limited surveys show a minimum of 20,000 individuals, but totals as high as 100,000 birds have been suggested. The southern population has been estimated at only 600 to 650 individuals, having declined substantially in the past century. The brolga is currently listed as threatened in Victoria.

The most significant threat to survival is loss and degradation of wetland habitat. Changes in hydrological processes and water quality through intensive livestock grazing, dams and drainage projects, water diversions and saltwater intrusion into coastal wetlands are having the greatest detrimental effects. Other hazards associated with increasing density of

human settlement include construction of power lines — with which flying birds collide — and wire fencing in breeding habitats that entraps flightless young, making them unable to follow parents to food and water sources. Chicks and eggs also suffer predation by introduced red foxes (*Vulpes vulpes*).

Both the Australian and New Guinea governments have signed the Ramsar Convention for wetland conservation. In addition, brolgas are legally protected in most Australian states but, although the birds use protected areas throughout their range, most brolga habitat is privately owned. Landowners have adopted few conservation measures; however, some habitat alterations, such as construction of water impoundments, have provided incidental benefit to the birds. In addition, various initiatives are under way to encourage voluntary beneficial wetland management. Nonetheless, more scientific research is required on this species, particularly in northern Australia and New Guinea.

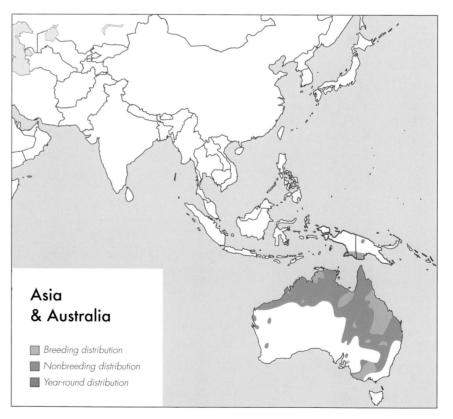

Asia
& Australia

■ Breeding distribution
■ Nonbreeding distribution
■ Year-round distribution

White-naped Crane
(Grus vipio)

White-naped cranes are found regularly in the company of other crane species, including Siberian, hooded and Eurasian cranes, which also winter at Poyang and Dongting lakes in China. White-naped cranes also share breeding areas with red-crowned cranes and demoiselle cranes in Mongolia and Russia.

Appearance

Height: 47–51 in (120–130 cm)
Weight: 10.3–14.3 lbs (4.7–6.5 kg)
Wingspan: 79–83 in (200–210 cm)
Adult dark gray and white striped neck and silvery gray wings. Bright red facial skin scattered with black bristly feathers. Bill yellowish gray. Males slightly larger than females. Juvenile gray with yellow- or brown-tipped feathers and fully feathered cinnamon-brown head. Downy chick tawny yellow with darker brown upper parts.

Distribution

White-naped cranes currently breed in scattered locations along rivers in southeastern Russia south to northeastern Mongolia and northeastern China. Their distribution was probably more widespread historically; however, accurate information is limited. It is likely that their range has contracted considerably since the early 1900s, when populations began to decline steadily due to multiple effects of human encroachment and warfare on nesting and wintering grounds and migratory stopover sites. No subspecies are recognized.

Seasonal Movements

White-naped cranes are migratory. Birds from eastern portions of their range migrate south to wintering grounds along the Korean peninsula. Several hundred individuals remain in or near the demilitarized zone between North and South Korea through the winter, although about 2,000 cranes continue south to the island of Kyushu in Japan, where they are sustained by artificial feeding programs. White-naped cranes nesting in the western part of the distribution migrate south across eastern China to winter in the Yangtze River valley.

Habitat

This species nests in open, shallow wetlands, in wet meadows in river valleys and along lake edges. They prefer wetlands adjacent to grasslands and croplands that offer additional foraging opportunities. During migration and on the wintering grounds, white-naped cranes also frequent fallow fields, marshes, rice paddies and estuarine mudflats.

Food Habits

When breeding, white-naped cranes feed on roots and tubers of sedges and other wetland plants, for which they dig vigorously. They also consume insects and small invertebrates. They eat more plant material, such as waste grain and seeds, during the nonbreeding season. Rice and cereal grains are used as supplemental winter food sources in Japan.

Behavior and Social Interactions

Mated pairs of white-naped cranes often vocalize in unison. The male gives a shrill, piercing *gar-oooo–gar-oooo–gar-oooo–gar-oooo* call of five to twelve syllables. The female utters *tuk-tuk–tuk-tuk–tuk-tuk–tuk-tuk*, with each syllable given immediately following the male's notes. Females frequently initiate the unison display. The birds may stand side by side or face each other as they vocalize. The male elevates his wings and both birds throw

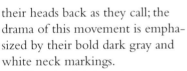

their heads back as they call; the drama of this movement is emphasized by their bold dark gray and white neck markings.

Dancing birds spring from the ground with wings spread, sometimes tossing objects. They often jump in synchrony and may call at times during their display. There usually is less head bobbing among white-naped cranes than in other species.

White-naped cranes are highly territorial when nesting and feeding. A male guards his family members by rushing at an intruder with his wings raised, neck arched and bill pointed forward and down. He may precede the attack with a bout of "false preening." Threatened intruders typically retreat before physical violence ensues.

Reproduction

White-naped cranes first breed at about three years of age, after which they typically nest each spring in April or May. Their nest is a broad platform of dried grasses and sedges with a central depression in which the eggs are placed. The female usually lays two buffy gray spotted eggs. Both males and females share incubation duties for 28 to 32 days. The parents rarely leave their nest unattended; the non-incubating bird typically roosts in shallow water nearby. The male assumes the primary role in defending the nest and chicks. Chicks learn to feed themselves within a few days of hatching but will readily take food offered to them by their parents. Although young fledge at 70 to 75 days of age, parents continue to care for their offspring through the winter. When the young birds are about 10 months old — just prior to the next breeding season — their parents drive them off.

Conservation Status

Population is declining.
World Conservation Union rating: vulnerable.

White-naped cranes currently number about 5,000 birds. Although accurate census information is lacking, some researchers have suggested that the world population may have been lower immediately following World War II and the Korean War due to extensive damage to habitat along migration routes and on wintering grounds. The eastern population has since been bolstered from several hundred individuals to over 2,000 through winter feeding programs in Japan. Unfortunately, the western population is now declining steadily.

Habitat degradation and loss is the most pressing concern to this species' survival, particularly for breeding areas along the Amur River, where wetlands and adjacent grassland are rapidly being converted to cropland. Intensification of agriculture brings additional threats, such as human disturbance, indiscriminate pesticide use, industrial pollution and modification to hydrological patterns. Moreover, the proposed dam projects in the Amur River basin and the completion of the Three Gorges Dam on the Yangtze River in China will likely destroy critical habitat areas. In China alone, damage to wintering habitat at Poyang and Dongting Lakes would threaten about 60 percent of the world population. Birds wintering in high concentrations at feeding stations in Izumi Crane Park in Japan are at increased risk of decimation due to disease outbreak.

Decline of white-naped crane populations is expected to continue.

Although they are legally protected, laws are rarely enforced. In the early 1990s, international workshops examined principal threats and identified habitat areas to be secured. Some breeding habitat has since been protected in Russia, China and Mongolia. Migrating birds stopping over in the Korean demilitarized zone are protected merely by circumstance; however, any change in political status of this region may prove detrimental. Some wintering habitat occurs in protected reserves in China, but most habitat management programs, if they exist, are underfunded and poorly organized. Only Izumi Crane Park in Kyushu provides true sanctuary for wintering populations.

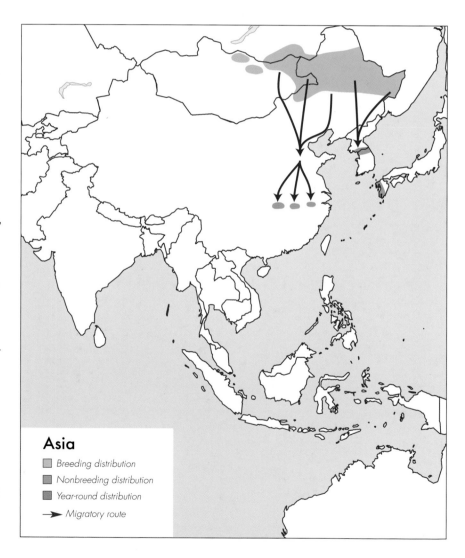

Asia
- ▪ Breeding distribution
- ▪ Nonbreeding distribution
- ▪ Year-round distribution
- → Migratory route

Eurasian Crane
(Grus grus)

The subtle beauty of the Eurasian crane has captured the hearts of artists for millennia. Its portrait has adorned the walls of caves in Spain, France and Great Britain for 12,000 years, and the tombs of the ancient Egyptians throughout that civilization's long history. Eurasian cranes, also called common cranes, have recently taken up residence in eastern England after an absence of 400 years.

Appearance
Height: 43–47 in (110–120 cm)
Weight: 9.9–13.4 lbs (4.5–6.1 kg)
Wingspan: 71–79 in (180–200 cm)
Adult slate gray with darker rump, back and nape. Head blackish except for bold white stripe behind eye and bare red crown. Bill yellowish brown. Very prominent "bustle" formed by elongated, drooping wing feathers. Juvenile gray with feathers tipped with cinnamon. Head fully feathered, but bustle lacking. Downy chick chestnut brown with pale underparts.

Distribution
Eurasian cranes have the widest breeding distribution of all crane species, ranging from Scandinavia and western Europe west across Eurasia to northeastern China, northern Mongolia and eastern Russia. They occupy most of their historical range; however, many localized extirpations have occurred throughout Europe and the United Kingdom. Seven breeding populations of Eurasian cranes have been identified, but no subspecies are currently recognized. The species was previously split into two subspecies based on plumage coloration but it has since been shown that these discrepancies were due to differences in feather-painting behavior among populations.

Seasonal Movements
Eurasian cranes are migratory. They travel along several major migration routes from northern Europe through western Europe to winter in France, the Iberian peninsula and Morocco; from northeastern Europe across central Europe south to eastern Africa; from eastern Europe and western and central Russia through the Balkans to wintering areas in eastern Africa and the Middle East; from western Siberia to India; and from central Siberia and northern China across China south to southern and eastern China. Eurasian cranes are rather gregarious during the nonbreeding season and sometimes form flocks as large as 400 members.

Habitat
Preferred nesting habitat includes shallow freshwater wetlands, such as open marshes, forested swamps, sedge meadows, lake edges and bogs. Although this species formerly used permanent densely vegetated marshes in Europe, it is now adapted to nesting in small artificial and restored wetlands where it had been previously extirpated. Drier habitats may be used in central Asia provided that some water is available. During migration, Eurasian cranes can be found in agricultural fields, shallow lakes and wet meadows. They winter in wetlands or other shallow waters but forage in fields, pastures and other upland areas.

Food Habits
Eurasian cranes consume a wide variety of plant materials, such as roots, tubers, shoots, berries, seeds, nuts, acorns and waste grain. They also eat worms, snails, insects, frogs, snakes, lizards, fishes and rodents, particularly in summer or when feeding chicks.

Behavior and Social Interactions
The bugling calls of the Eurasian crane are high-pitched, penetrating and loud, frequently in excess of 100 decibels. They often vocalize in unison

in response to threat. Typically, females initiate a unison call with a long scream that is followed by several shorter, regularly spaced notes. Males answer with a single long note. They variably raise their wings to expose dark primary feathers when calling, perhaps accordingly to the degree of threat or enthusiasm.

Dancing Eurasian cranes display bobs, pirouettes and leaps, often accompanied by tossing vegetation into the air. An aggressive territorial male crane will also perform a "parade march" with legs lifted high, or "false preen" with his ruffled wings held away from his body. He frequently directs his bare red crown to an opponent during these displays. Eurasian cranes will thwart the attacks of large predators, such as sea eagles (*Haliaeetus albicilla*), by leaping into the air and striking with their bill or claws. These counterattacks may also occur while in flight.

Breeding birds often paint the feathers of their upper body with reddish brown mud or decaying vegetation, a behavior that they share with sandhill cranes. Feather painting is thought to camouflage these two primarily gray species when nesting in a dead-grass habitat.

Reproduction

Eurasian cranes breed for the first time at four or six years of age. They nest in the spring; most eggs are laid in May. Mated pairs often use the same nesting site from year to year in a well-defended territory that varies in size from 125 to 1,235 acres (50 to 500 hectares) or more. Nests are typically built from wetland vegetation in shallow water near trees. The female lays two spotted olive-brown eggs. Both parents participate in incubation for about 30 days; however, the female typically makes the greatest contribution. The nonincubating bird roosts some distance from the nest, but will perform a showy distraction display — drooping one wing and limping — for potentially dangerous predators. Less threatening intruders are attacked directly by bill stabbing, kicking and wing beating. Downy young can leave the nest within a few hours of hatching, and run and swim well after a day. They respond to threats by squatting and freezing. Young fledge at age 65 to 70 days, but remain in the care of their parents for many months. Annual reproduction success varies somewhat. Studies have reported nesting success rates of 50 to 77 percent, with twins being produced 20 to 30 percent of the time, a figure considerably higher than for most other crane species.

Conservation Status

Population is generally stable; western populations increasing, many central and eastern populations decreasing. World Conservation Union rating: least concern.

With an estimated total population of about 220,000 to 250,000 individuals, the Eurasian crane is not globally threatened. However, some populations, particularly those in the east, are declining. In addition, the lack of comprehensive population surveys throughout much of this species' range has brought into question the reliability of past estimated numbers.

Principal threats to Eurasian crane populations are loss and degradation of wetlands. This has been most frequently associated with human development, and includes draining or damming wetlands or river systems for agricultural expansion and urbanization. Hunting and egg collecting are also of concern, and likely contributed significantly to their extirpation from England and southern Europe.

Populations in western and central Europe have benefited considerably from conservation measures such as legal protection, monitoring programs, wetland creation and restoration initiatives, and protection of existing habitat in breeding, wintering and migratory staging areas, which accounts for some local population increases. Unfortunately, theses activities are rare or absent in the remainder of the Eurasian crane's distribution and, although international cooperation for crane conservation is now dawning in eastern Europe, not all range countries have signed the Ramsar Convention. Recently, Eurasian cranes have been the subject of several scientific research programs in China, Russia, India, Pakistan and Israel; perhaps the results of these studies will encourage greater participation in conservation programs. Eurasian cranes frequently share habitat with other crane species; any conservation efforts would likely benefit all.

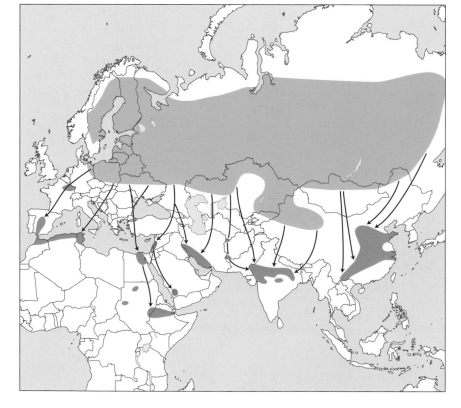

■ Breeding distribution
■ Nonbreeding distribution
■ Year-round distribution
➡ Migratory route

Hooded Crane
(Grus monacha)

The hooded crane is a rare species with a notably remote breeding distribution. Consequently the first active nest was not discovered until 1974, and a small breeding population in China eluded researchers until the 1990s. Lack of human disturbance on breeding grounds has been beneficial to this vulnerable species; unfortunately, the same cannot be said of its wintering distribution, where high-density human populations have modified its traditional habitat considerably.

Appearance
Height: 39–43 in (100–110 cm)
Weight: 7.3–10.6 lbs (3.3–4.8 kg)
Wingspan: 63–71 in (160–180 cm)
Adult slaty gray with white head and neck. Bare red crown thatched with black bristles. Flight feathers black; inner wing feathers elongated below knees. Males slightly larger than females. Juvenile grayish brown with black and white feathered crown. Downy chick rusty brown with gold-tipped feathers.

Distribution
Hooded cranes breed primarily in eastern Siberia from Lake Baikal and southern Yakutia to the lower Amur River basin. There are limited nesting areas in northwestern Manchuria in northern China; however, survey information is lacking. Nomadic non-breeding flocks also occur along the Russia-Mongolia-China border during summer. Little is known about this species' historical distribution, particularly in breeding areas. Also, many records of hooded cranes in summer may represent birds that are not nesting. No subspecies are recognized.

Seasonal Movements
This migratory species travels to winter destinations along two disparate routes. Most travel south through eastern Inner Mongolia and northeastern China down the Korean peninsula to Japan. About 8,000 individuals (80 percent of the total population) winter at the Izumi Feeding Station on Kyushu Island. Other destinations in Japan include Yashino on Honshu Island. Hooded cranes also migrate through northeastern China south to wintering grounds along the Yangtze River. The largest concentration of wintering birds in China (about 300) occurs at Shengjin Lake in Anhui Province.

Habitat
Hooded cranes nest in isolated wet sphagnum bogs in taiga habitat that is not too open or too densely forested. Adults with young also feed in grassy meadows along small rivers. Non-breeding birds may be found in shallow open wetlands, agricultural fields and grasslands in summer. Cranes wintering in China frequent shallow lakes, mudflats, grassy marshes and agricultural fields. They occur primarily in grain fields, rice paddies and artificial feeding stations in Japan during winter.

Food Habits
Dietary choices on breeding grounds are poorly known. Plant material, such as aquatic vegetation and berries, is likely consumed in greatest quantities; however, insects, frogs and salamanders are also eaten. In winter, hooded cranes eat rice, wheat and other grains, particularly at artificial feeding stations.

Behavior and Social Interactions
The voice and dancing displays of the hooded crane resemble those of other *Grus* species, such as the Eurasian, whooping and red-crowned cranes,

although they have rarely been described. Pairs give loud, high-pitched unison calls. The male raises his wings to form a plume over his back while extending his head up. His disyllabllic *gar-raw* call is answered by one long and one short note by the female. Males tend to be more vocal than females, although females typically begin a bout of unison calling.

Hooded cranes are faithful to the same nest site from year to year and may even renovate and reuse old nests; shell fragments resulting from two or three nesting events have been found. Hooded and Eurasian cranes share nesting habitat in parts of their range, and often associate with each other during migration. Consequently, mixed pairing resulting in hybrid offspring has been observed on occasion.

Reproduction

Hooded cranes begin to reproduce at age three or older. Nesting occurs in the spring, usually April or May. Nests are built from mosses, peat and sedges, often with stalks, leaves and branches placed at the top. The female typically lays two eggs, which are incubated alternately by both parents for 27 to 30 days. The pair is quiet and secretive during incubation. When arriving or leaving the nest site, they remain close to the ground as they skulk through the vegetation. Chicks are active within three days of hatching but stay near the nest. By age one week, however, they are able to follow their parents a mile (2 km) or more to feeding areas. The chicks fledge about 75 days of age; migration occurs shortly after.

Conservation Status

Population is declining.
World Conservation Union rating: vulnerable.

The global estimate of hooded cranes may be as few as 9,400 individuals. Nonetheless this species is, perhaps, less prone to extinction than other vulnerable cranes because the remoteness of their breeding sites has provided some protection from habitat loss and persecution. In addition, woodland bogs that they frequent are less likely to be converted to agriculture, and logging activities in these areas typically take place in winter when the birds are not present.

Hence, the greatest threats to hooded crane survival occur on wintering grounds where human densities are high. Winter habitats in China are being lost rapidly to conversion to agriculture. As preferred habitat dwindles, cranes often resort to using rice paddies and crop fields as winter roosts and feeding sites. There, they are perceived as pests by farmers and are shot or intentionally poisoned as they feed on tainted grain that they are offered. Other threats to habitat in China and Korea include changes to hydrological

patterns associated with dams and highway construction.

Although hooded cranes are held in high esteem in Japan, they are prone, nonetheless, to numerous threats associated with human development. For example, many traditional agricultural activities that favored cranes are now occurring in large enclosed greenhouses. In 1945, an airport was built in Izumi beside a primary crane-roosting area, and the birds were disturbed regularly to avoid collisions with aircraft. Also, cranes

continue to perish in developed areas when they strike utility lines. Artificial feeding stations and intensive habitat management in Japan have been instrumental in maintaining and encouraging growth in hooded crane populations; however, the high density of cranes wintering in these areas favors decimation through disease outbreak. Conservation efforts in China and Korea have been lacking; a few protected areas have been established with varying degrees of success.

Asia
- Breeding distribution
- Nonbreeding distribution
- Year-round distribution
- → Migratory route

Whooping Crane
(*Grus americana*)

The whooping crane is the only threatened or near-threatened crane species with a general upward trend in population size. Whoopers are so named because of their loud, trumpeting call that is produced by their extensively coiled trachea, or windpipe.

Appearance
Height: 51–63 in (130–160 cm)
Weight: 9.9–18.7 lbs (4.5–8.5 kg)
Wingspan: 79–91 in (200–230 cm)
Adult white with black primary wing feathers and black facial stripe. Bare red skin on crown sparsely thatched with black bristles. Eyes and bill yellowish. Males slightly larger than females. Whitish juvenile heavily mottled with cinnamon; feathered head pale, warm brown. Cinnamon-brown chick has grayish underparts. Blue eyes become aqua at three months and yellow at six months of age.

Distribution
Whooping cranes occur in three isolated breeding populations but no subspecies are recognized. The only natural, self-sustaining flock nests in Wood Buffalo National Park in the Canadian Northwest Territories. A second migratory population of experimentally reintroduced birds summers in Necedah National Wildlife Refuge in central Wisconsin. An experimental nonmigratory flock has also been established in Kissimmee Prairie in south-central Florida. Fossil evidence dating back several million years indicates that whooping cranes once had a much larger distribution in North America, extending from central Canada to Mexico, and from Utah east to the Atlantic Ocean. Recent extirpations include a wild, nonmigratory population that occurred in southern Louisiana and a small experimental flock introduced into southeastern Idaho.

Seasonal Movements
There are two migratory populations of whooping cranes. The Wood Buffalo National Park flock migrates south across the central plains of North America to winter in Aransas National Wildlife Refuge on the Texas Gulf Coast. An important spring and fall migratory staging area occurs on the Platte River in Nebraska. The experimental migratory population travels south from Wisconsin to the Chassahowitzka National Wildlife Refuge on the Gulf Coast of Florida.

Habitat
Migratory whooping cranes nest in wetlands with bulrushes, cattails and sedge vegetation, such as small ponds, marshes and muskeg. Traditional nesting habitats also included flooded prairie potholes. They use many wet habitat types during migration, including riparian marshes, reservoir margins, submerged sandbars and agricultural fields. Wintering migratory populations frequent brackish bays, salt flats and coastal marshes. Nonmigratory cranes also use wet meadows, savannas and lake margins in Florida.

Food Habits
On the breeding grounds, whooping cranes consume primarily animal foods, including mollusks, crustaceans, aquatic insects, minnows, frogs and snakes. They occasionally eat berries. Migrating birds also eat waste grain in harvested fields and plant tubers. Crabs and clams are important food choices for birds wintering in coastal areas. Cranes feeding in upland areas in winter also eat acorns, snails, small rodents, grasshoppers and snakes.

Behavior and Social Interactions
The whooping crane's voice is loud and trumpeting, and can be heard over great distance. When alarmed, individuals produce a penetrating single *kerloo*!, following which the bird may prepare to attack or take flight. Pairs also give a coordinated bugled unison call. The male gives a low-pitched *kronk–kronk–kronk–kronk–kronk* while he raises his head and lowers his black wing tips. Syllables of the female's call often change in frequency throughout, which produces a glissando effect. Females sometimes give two or three short notes in response to each note given by the male.

Whooping cranes often dance as part of their courtship behavior. The dance often begins with bowing and wing flapping, which is followed by repeated leaps into the air on stiffened legs with the bill pointing skyward. Wings may be flapped during the leaps to display the black primary wing feathers that contrast starkly with the bird's white plumage. The frequency of leaping may increase until both birds are bowing and jumping two or three times in succession.

Male whooping cranes are the primary defenders of their family and nesting territories. When they encounter a potential threat, they perform a strutting display in which they turn their bodies sideways to the intruder and walk slowly with their feet lifted high. If the bird is strongly committed to defending his site, he will lower his head until his bill touches the ground and emit a low, menacing growl.

Reproduction

Whooping cranes begin breeding at about four or five years of age. Breeding occurs in spring, usually late April to May. Pairs mate for life and are typically faithful to the breeding territory that they used in previous years; territories measure 0.75 to 1.8 miles (1.2 to 2.9 km) in diameter. Nests are constructed in shallow water from aquatic vegetation such as bulrushes and sedges. The female usually lays two olive-brown eggs that are blotched with brown; both parents equally share incubation duties, which last 30 to 35 days. The nonincubating parent guards the nest from a distance and chases away potential predators. If a particularly large predator approaches, the incubating parent may leave the nest and lead the predator away using a wing-dragging distraction display. Downy chicks can walk or swim within a few hours of hatching. Parents readily offer food to a chick when it pecks at their lowered bill. Adults typically mash food items before presenting them to their young. Parental care is attentive and extended; juveniles six to nine months of age have been observed successfully begging food from their parents. Young

fledge at 80 to 90 days of age. Whooping cranes migrate in family groups and do not entirely leave their parents' care until the following spring.

Conservation Status

Population is increasing.
World Conservation Union rating: endangered.
ESA rating: endangered.

The recovery of the whooping crane from the brink of extinction has been long and arduous. The wild world population is about 350 individuals, making them easily the rarest crane species. Although whoopers may never have been very abundant in historical times, they certainly would have numbered in the many thousands. Unfortunately, their populations declined precipitously in the late 1800s and early 1900s as European settlement of North America became widespread. Principal causes of decline were hunting, egg and specimen collecting, conversion of preferred wetland habitats to agriculture and other forms of human disturbance. The one remaining wild migratory population

had reached a staggering low of about 15 birds by 1940.

In the 60 years that followed, enforceable legal protection, preservation of critical habitat and international cooperation in conservation initiatives fostered a steady increase in whooping crane numbers. Captive breeding programs have been used since the 1960s to bolster numbers and to provide individuals for reintroduction into experimental populations in Idaho, Wisconsin and Florida. Attempts to cross-foster whooping cranes to sandhill cranes in order to induce migration were disappointing. In 2001, however, migratory behavior was successfully established in seven captive-bred whoopers as they were led from Wisconsin to Florida by ultralight aircraft. Six years later, this group of cranes, and others escorted to wintering sites in subsequent autumns, faithfully retrace their route back to summering habitat in Wisconsin every spring. Demonstration of pair formation and nesting behavior among this reintroduced flock, now 66 birds strong, provides considerable reason for celebration.

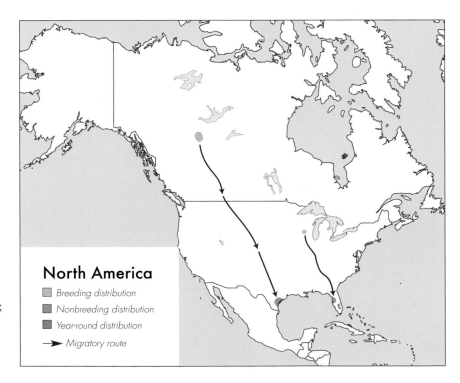

North America
- Breeding distribution
- Nonbreeding distribution
- Year-round distribution
- → Migratory route

Black-necked Crane
(Grus nigricollis)

Black-necked cranes occur at high elevations in remote regions of the Himalayas and neighboring mountain ranges. Consequently, in 1876, they became the last crane species to be described by European biologists. This does not suggest that this symbol of luck and happiness was unknown prior to this; on the contrary, its lovely image has graced the pages of historical books and the walls of Eastern temples for centuries.

Appearance
Height: 43–47 in (110–120 cm)
Weight: 11–15.4 lbs (5.0–7.0 kg)
Wingspan: 71–79 in (180–200 cm)
Adult ashy gray; head and neck black except for small whitish spots extending back from eyes, and red crown and lores (area between eyes and bill). Crown skin sparsely covered with black bristles. Eyes yellow. Flight feathers black. Males slightly larger than females. Juvenile yellowish gray with black and white neck and brownish-feathered crown. Downy chick brownish.

Distribution
Black-necked cranes occur on the Qinghai-Tibetan Plateau in western and central China, with breeding confirmed in Qinghai, Tibet, Sichuan, Gansu and Xinjiang provinces. Two very small, isolated populations also breed in eastern Ladakh and northern Sikkim in India. The species' historical distribution is poorly known, but it was likely more widespread than it is today. For example, nesting has occurred in Lhasa, Tibet. No subspecies are recognized.

Seasonal Movements
This species travels limited distances from high-elevation breeding areas (9,675 to 16,000 feet/2,950 to 4,900 m) to winter at lower elevations (6,250 to 13,000 feet/1,900 to 3,950 m). Wintering birds have been recorded in south-central Tibet in the Nyang, Lhasa and Pengbo River valleys and along the Yarlang Zsangbo; in China at lower elevations of the Qinghai-Tibetan and Yunnan-Guizhou plateaus; and in Butan and Arunachal, India. Historically, wintering birds were recorded in northern Vietnam.

Habitat
Black-necked cranes nest in high-elevation freshwater wetland habitats, such as grassy marshlands, small ponds in sedge-bog meadows and lakeshore or riparian marshes. In winter, they prefer open agricultural valleys and deserted rice paddies, where they feed mainly on waste grains. Black-necked cranes are less aquatic in winter than many other species. They are quite tolerant of human activities and regularly occur around settlements and grazing livestock.

Food Habits
During the breeding season, black-necked cranes consume fish, snails, algae, moss and submerged aquatic plants. They rely on waste grain in winter but also eat plant roots and tubers, earthworms and hibernating frogs that they excavate from boggy areas adjacent to agricultural lands.

Behavior and Social Interactions
Black-necked crane calls are high-pitched and penetrating. Mated pairs call in unison, typically while standing erect with their heads thrown back and bills pointing up. Females usually initiate the display, although sometimes the male will walk around the female while holding his wings outstretched, to lead her to dance with him. Dancing black-necked cranes leap, prance, bow and wave their heads.

The male is very attentive to intruders and typically takes the lead in family and territory defense. Before a male black-necked crane attacks a potential threat, he flaps his wings and raises them above his back, bends low to the ground and runs in circles to demonstrate his aggression. Nesting black-necked cranes are often seen in association with other wetland bird species, such as the ruddy shelduck (*Tadorna ferruginea*), and may use these attentive species to cue them to potential danger.

Reproduction
Black-necked cranes are spring breeders; most eggs are laid in May and June. They build their nest from a large pile of aquatic vegetation in a

shallow marsh or lake border. Nests are often surrounded by water to protect them from terrestrial predators. The female lays one or two eggs that are dark olive and strongly blotched with brown. Both parents alternate incubating the eggs for 31 to 33 days. The nonincubating bird typically forages within 650 to 1,000 feet (200 to 300 m) of the nest but remains watchful for evidence of disturbance. Males usually incubate at night. Chicks can move around the nest area within two days of hatching. By the third day, the parents take them on short feeding forays in grassy or sandy areas. They encourage foraging behavior in chicks by placing food on the ground in front of them. The family moves to more distant, richer feeding areas when the chick is about 10 days old. Young fledge at about 90 days of age but stay with their parents throughout their first winter.

Conservation Status

Population is declining.
World Conservation Union rating: vulnerable.

Black-necked cranes, numbering only 6,000 to 8,000 individuals, are the fourth-rarest crane species. Perhaps the greatest population declines occurred in the mid-1900s and, although evidence is sketchy, many local reports suggest that numbers have stabilized somewhat since the 1970s. Unfortunately, remnant populations occurring in much of their breeding range are now hazardously low.

Loss and degradation of habitat are the principal threat to black-necked cranes, particularly on wintering grounds, where increasing human populations modify prime habitat through damming or draining watercourses, extraction of water for irrigation, flood-control projects and conversion of land to cropland or tree plantations. In Yunnan and Guizhou, wetlands are being converted to deep fishing ponds that are unsuitable for cranes. Other threats include peat

mining for fuel, proposed hydroelectric projects, deforestation, soil erosion, siltation of wetlands and industrial pollution. Furthermore, black-necked cranes wintering in Tibet have been affected by changes to traditional agricultural practices that have caused a substantial reduction in the availability of waste barley and wheat — two primary winter foods for this species. This has led to more crop damage (potatoes, maize and carrots) by hungry cranes that, in turn, causes increased persecution by farmers.

Fortunately, black-necked cranes are legally protected throughout their range. More importantly, this species is revered where Buddhism prevails because religious beliefs prevent hunting of wildlife. High fines or imprison-

ment is imposed for illegal hunting and egg collecting in Tibet. In Yunnan, farmers who provide aid to a sick crane are rewarded. Black-necked cranes are the provincial bird of Qinghai and the symbol of Bhutan's Royal Society for the Preservation of Nature.

International cooperation has promoted conservation efforts such as identification and subsequent establishment of some protected areas, watershed management programs, reforestation initiatives to reduce siltation in wetlands and scientific monitoring and research. Black-necked cranes breed readily in captivity but to date recovery efforts have been gaining ground without immediate need for reintroduction programs.

Breeding distribution
Nonbreeding distribution
Year-round distribution

Red-Crowned Crane
(Grus japonensis)

The people of Japan have long been devoted to this beautiful bird. Red-crowned crane pairs remain bound to one another through life, and are thus bestowed with traditional reverence as symbols of fidelity and love. Japanese legends also recognize these long-lived birds as symbols of longevity; ancient tales chronicle the millennial life of mythical red-crowned cranes. More recently, folded-paper origami cranes have come to embody the hope for peace.

Appearance
Height: 55–63 in (140–160 cm)
Weight: 15.4–22 lbs (7.0–10.0 kg)
Wingspan: 87–98 in (220–250 cm)
Adult white with black (male) or pearly gray (female) cheeks. Only species with white primary wing feathers; secondary and tertiary feathers black. Bare red skin on forehead and crown; bold white band extends from below eyes down back of head and neck. Bill olive green. Male slightly larger than female. Juvenile plumage rusty white with brownish gray neck and secondaries. White primaries tipped with black; fully white primaries acquired at two years of age. Downy chick tawny or cinnamon brown.

Distribution
There are two primary populations of red-crowned cranes. A migratory population — about two-thirds of the entire species — breeds in Inner Mongolia and Heilongjiang, Jilin and Liaoning provinces in northeastern China and adjacent regions in southeastern Russia. Remaining birds are nonmigratory residents on the Japanese island of Hokkaido. These two populations exhibit some differences in vocalizations, plumage and egg size; however, they are no longer considered subspecies. Red-crowned cranes undoubtedly had a much larger breeding range historically, but records are poor and there is some confusion whether birds sighted extralimitally in China, Korea and Japan are breeding, migrant or wintering birds. Regardless, this species' current distribution represents only fragments of its former range.

Seasonal Movements
Migrant red-crowned cranes travel southeast across northeastern China to form three or four wintering subpopulations. Birds from the eastern breeding range winter primarily in or near the demilitarized zone on the Korean peninsula. Western birds fly south across the north China Sea to wintering grounds in coastal Jaingsu Province, China. The nonmigratory population in Japan is mostly sedentary; however, some family groups travel about 90 miles (150 km) to the Kushiro District of Hokkaido.

Habitat
Red-crowned cranes are among the most aquatic cranes. They prefer to nest and feed in relatively deep-water marshes with standing dead vegetation, such as cattails, reeds and sedges. They also use bogs and wet meadows. In winter, this species occurs in coastal salt marshes, mudflats, rivers, freshwater marshes, rice paddies and cultivated fields. They are well adapted to cold temperatures.

Food Habits
Preferred foods include insects and aquatic invertebrates, fishes, frogs and small rodents. They also eat reeds, grasses and berries. Waste rice and other grains are consumed in winter. Populations in Hokkaido rely on corn, cereal grains and fish provided at artificial feeding stations.

Behavior and Social Interactions
The voice of the red-crowned crane is loud and high-pitched. The male gives a penetrating, whooping, rattling *kar-r-r-o-o-o* call singly in winter or in a multisyllabic unison call throughout the year. The female responds with her *tuk–tuk–tuk–tuk–tuk–tuk*. During the unison call, male red-crowned cranes hold their secondary flight feathers high above their back and point their bill skyward. The female also points her bill up, but keeps her wings somewhat lower. The male may strut around the female as he calls.

Red-crowned cranes of nearly all ages dance, particularly in late winter and early spring. Mated pairs often perform simultaneously as they face one another. Their dance includes much bowing and jumping with dangling legs, usually in the direction that the bird is already facing.

These cranes vigorously defend large nesting territories that measure about 0.4 to 4.5 square miles (1 to 12 square km). Other cranes are evicted rapidly from territories, as are potential predators. An aggressive male first displays his intentions by "false preening," then points his bill downward while facing his opponent. If this threat is not sufficient to drive away an intruder, he will charge and attack using his stabbing bill, slashing claws and wildly beating wings as weapons.

Reproduction

Red-crowned cranes typically begin breeding at about four years of age. From April to June, they select a nest site on wet ground or in standing water up to 18 inches (45 cm) deep in areas where emergent vegetation is 12 to 80 inches (30 to 200 cm) tall. Their nest is made of reeds and grass. The female typically lays two eggs; color and size of eggs are variable. Both parents incubate for 29 to 34 days. Incubating parents are diligent, particularly early in the nesting period when cold overnight temperatures can freeze unattended eggs. The male is the primary defender of the nest, and is ever watchful when he is not on incubation duty. If food is plentiful near the nest, parents and chicks will remain in the area for one to two weeks. After this, they often move to drier habitats or marsh borders to feed. Both parents are attentive to their young and may continue to feed them for many weeks or months. Chicks fledge at about 95 days of age but stay with their parents for almost a year. Annual reproductive success among fully protected birds breeding in Japan appears to be 12 to 15 percent; success for other populations is less well known.

Conservation Status

Population is declining.
World Conservation Union rating: endangered.

The red-crowned crane is the second-rarest crane species, with a total wild population of about 2,000 birds. Although this number seems startlingly close to extinction, it represents an increase from a historical low following World War II. Nonetheless, this species' limited recovery has occurred in only part of its distribution, largely as a result of artificial feeding stations.

Cranes continue to be seriously threatened by loss of wetland habitat throughout their range, even in Japan, where they are held in high esteem. Agricultural expansion, river channelization, deforestation and road building have destroyed many nesting areas. In China, preferred wetland habitats are being converted rapidly to agricultural use; in northern China, an estimated 90 percent have been drained since the 1950s. Between 1979 and 1984 alone, two-thirds of the suitable breeding habitat on the Dulu River was lost. Other threats in China include oil drilling, cutting and burning of reeds in wetlands, and water control and diversion projects. In Korea, red-crowned cranes currently winter in or near the demilitarized area that functions unofficially as a protected area, but any change in status would undoubtedly threaten these populations. Agricultural intensification, particularly livestock grazing, is also reducing

crane breeding habitat in Russia. Development of dams along the Amur and Yangtze rivers will be particularly devastating to preferred habitat. Although red-crowned cranes are protected by law, hunting and egg collecting and poisoning occur regularly in Russia and China.

Red-crowned cranes have benefited under the Ramsar Convention for Wetland Conservation, which has been signed by China, Japan and Russia. This international agreement has encouraged the exchange of summer and winter survey information and fostered ongoing scientific research. Wetland reserves in China now support more than 500 breeding red-crowned cranes; other protected areas have been proposed. Some migratory stopover and wintering sites in China and Korea have also been designated as protected areas. Winter feeding stations in Japan have increased winter survivorship, and installation of conspicuous markers on utility lines have reduced mortality from deadly collisions. Nonetheless, high densities of birds at feeding stations increase the risk of catastrophic loss due to disease outbreak. Red-crowned cranes have been breeding in captivity for centuries; ongoing programs to maintain genetic diversity and some reintroduction projects are underway.

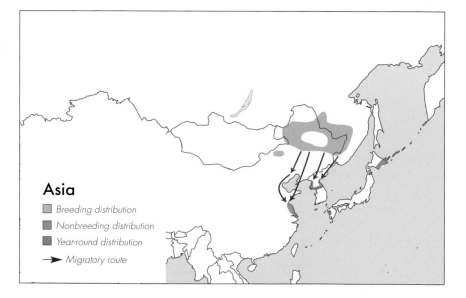

Asia
- ■ Breeding distribution
- ■ Nonbreeding distribution
- ■ Year-round distribution
- → Migratory route

Appendix I

World Conservation Union Ratings

Species	Total Population	Population Trend	IUCN Rating
Grey crowned crane	90,000	Declining	Least concern
Black crowned crane	65,000	Declining	Near threatened
Demoiselle crane	200,0000 to 240,000	Stable with local declines	Least concern
Blue crane	21,000	Declining	Vulnerable
Wattled crane	13,000 to 15,000	Declining	Vulnerable
Siberian crane	3,000	Declining	Critically Endangered
Sandhill crane	500,000	Stable with local declines	Two subspecies endangered
Sarus crane	13,000 to 15,000	Declining	Vulnerable
Brolga crane	20,000 to 100,000	Stable with local declines	Least concern
White-naped crane	5,000	Declining	Vulnerable
Eurasian crane	220,000 to 250,000	Stable with local declines	Least concern
Hooded crane	9,400	Declining	Vulnerable
Whooping crane	350	Increasing	Endangered
Black-necked crane	6,000 to 8,000	Declining	Vulnerable
Red-crowned crane	2,000	Declining	Endangered

Appendix II

Wild Whooping Crane Peak Winter Populations from 1938/1939 to 1970/1971

Winter	Wood Buffalo–Aransas			Louisiana	Rocky Mountain			Kissimmee Prairie			Necedah–Chassahowitzka			Total
	Adult	Young	Subtotal	Adult	Adult	Young	Subtotal	Adult	Young	Subtotal	Adult	Young	Subtotal	
1938/1939	14	4	18	11										29
1939/1940	15	7	22	13										35
1940/1941	21	5	26	6										32
1941/1942	14	2	16	6										22
1942/1943	15	4	19	5										24
1943/1944	16	5	21	4										25
1944/1945	15	3	18	3										21
1945/1946	18	4	22	2										24
1946/1947	22	3	25	2										27
1947/1948	25	6	31	1										32
1948/1949	27	3	30	1										31
1949/1950	30	4	34	1										35
1950/1951	26	5	31	0										31
1951/1952	20	5	25											25
1952/1953	19	2	21											21
1953/1954	21	3	24											24
1954/1955	21	0	21											21
1955/1956	20	8	28											28
1956/1957	22	2	24											24
1957/1958	22	4	26											26
1958/1959	23	9	32											32
1959/1960	31	2	33											33
1960/1961	30	6	36											36
1961/1962	34	5	39											39
1962/1963	32	0	32											32
1963/1964	26	7	33											33
1964/1965	32	10	42											42
1965/1966	36	8	44											44
1966/1967	38	5	43											43
1967/1968	39	9	48											48
1968/1969	44	6	50											50
1969/1970	48	8	56											56
1970/1971	51	6	57											57

Appendix II (cont'd.)

Wild Whooping Crane Peak Winter Populations from 1971/1972 to 2006/2007

Winter	Wood Buffalo–Aransas			Louisiana	Rocky Mountain			Kissimmee Prairie			Necedah–Chassahowitzka			Total
	Adult	Young	Subtotal	Adult	Adult	Young	Subtotal	Adult	Young	Subtotal	Adult	Young	Subtotal	
1971/1972	54	5	59											59
1972/1973	46	5	51											51
1973/1974	47	2	49											49
1974/1975	47	2	49											49
1975/1976	49	8	57			4	4							61
1976/1977	57	12	69		3	3	6							75
1977/1978	62	10	72		6	2	8							80
1978/1979	68	7	75		6	3	9							84
1979/1980	70	6	76		8	7	15							91
1980/1981	72	6	78		15	5	20							98
1981/1982	71	2	73		13	0	13							86
1982/1983	67	6	73		10	4	14							87
1983/1984	68	7	75		13	17	30							105
1984/1985	71	15	86		21	12	33							119
1985/1986	81	16	97		27	4	31							128
1986/1987	89	21	110		20	1	21							131
1987/1988	109	25	134		16	0	16							150
1988/1989	119	19	138		14	0	14							152
1989/1990	126	20	146		13	0	13							159
1990/1991	133	13	146		13	0	13							159
1991/1992	124	8	132		12	0	12							144
1992/1993	121	15	136		9	0	9							145
1993/1994	127	16	143		8	1	9	8		8				160
1994/1995	125	8	133		4	0	4	16		16				153
1995/1996	130	28	158		3	0	3	25		25				186
1996/1997	144	16	160		3	0	3	56		56				219
1997/1998	152	30	182		3	3	6	60		60				248
1998/1999	165	18	183		4	0	4	57		57				244
1999/2000	171	17	188		2		2	65		65				255
2000/2001	171	9	180		2		2	74		74				256
2001/2002	161	15	176		1		1	87		87		6	6	270
2002/2003	169	16	185		0		0	85	1	86	5	16	21	292
2003/2004	169	25	194					85	2	87	20	16	36	317
2004/2005	182	33	215					61	5	66	32	13	45	326
2005/2006	189	25	214					58	0	58	41	23	64	336
2006/2007	191	45	236					41	4	45	55	4	59	340

References

Allen, R.P. 1953. *The Whooping Crane. National Audubon Society Research Report No. 3.* National Audubon Society, New York, New York.

Allen, R.P. 1956. *A Report on the Whooping Crane's Northern Breeding Grounds. National Audubon Society Research Report No. 3 Supplement.* National Audubon Society, New York, New York.

Allen, R.P. 1957. *On the Trail of Vanishing Birds.* McGraw-Hill, New York, New York.

Archibald, G.W., and C. Meine. 1995. *Family Gruidae.* Pages 60-89 in *Handbook of the Birds of the World,* vol. 2. Lynx Edicions, Barcelona.

Canadian Wildlife Service and U.S. Fish and Wildlife Service. 2005. *Draft International Recovery Plan for the Whooping Crane.* Recovery of Nationally Endangered Wildlife (RENEW), Ottawa, and U.S. Fish and Wildlife Service, Albuquerque, New Mexico.

Coerr, E. 1977. *Sadako and the Thousand Paper Cranes.* Puffin Books, New York, New York.

Doughty, R.W. 1989. *Return of the Whooping Crane.* University of Texas Press, Austin, Texas.

Ellis, D.H. (ed.). 2001. *Proceedings of the Eighth North American Crane Workshop.* North American Crane Working Group, Seattle, Washington.

Ellis, D.H. 2001. *Wings Across the Desert.* Hancock House Publishers, Blaine, Washington.

Ellis, D.H., G.F. Gee, and C.M. Mirande (eds.). 1996. *Cranes: Their Biology, Husbandry, and Conservation.* Hancock House Publishers, Blaine, Washington.

Johnsgard, P.A. 1983. *Cranes of the World.* Indiana University Press. Bloomington, Indiana.

Johnsgard, P.A. 1991. *Crane Music: A Natural History of North American Cranes.* University of Nebraska Press, Lincoln, Nebraska.

Katz, B. 1993. *So Cranes May Dance: A Rescue from the Brink of Extinction.* Chicago Review Press, Chicago, Illinois.

Leopold, A. 1949. *A Sand County Almanac and Sketches from Here to There.* Oxford University Press, New York, New York.

Lewis, J.C. 1999. *Whooping Crane* (Grus americana). No. 153 in *The Birds of North America* (A. Poole and F. Gill, eds.). The Birds of North America Inc., Philadelphia, Pennsylvania.

Lishman, W. 1995. *Father Goose.* Little, Brown and Company, Toronto, Ontario.

Matthiessen, P. 2001. *The Birds of Heaven: Travels with Cranes.* Greystone Books, Vancouver, British Columbia.

McCoy, J.H. 1966. *The Hunt for the Whooping Crane.* Lothrop, Lee and Shepard, Inc., New York, New York.

McNulty, F. 1966. *The Whooping Crane: The Bird That Defies Extinction.* Clarke, Irwin and Company Limited, Toronto, Ontario.

Meine, C.D., and G.W. Archibald (eds.). 1996. *The Cranes: Status Survey and Conservation Action Plan.* IUCN, Gland, Switzerland, and Cambridge, United Kingdom.

Pratt, J.J. 1996. *The Whooping Crane: North America's Symbol of Conservation.* Castle Rock Publishing, Prescott, Arizona.

Price, A.L. 2001. *Cranes: The Noblest Flyers.* La Alameda Press, Albuquerque, New Mexico.

Scott, D.H. 1990. *A Flight of Cranes: Stories and Poems from Around the World about Cranes.* Denvil Press, West Sussex, United Kingdom.

Stahlecker, D.W. (ed.). 1992. *Proceedings of the Sixth North American Crane Workshop.* North American Crane Working Group, Grand Island, Nebraska.

Stehn, T.V. 2001–06. *Whooping Crane Recovery Activities: Semiannual Updates.* US Fish and Wildlife Service unpublished reports, Aransas, Texas.

Temple, S.A. (ed.). 1978. *Endangered Birds: Management Techniques for Preserving Threatened Species.* University of Wisconsin Press, Madison, Wisconsin.

United States Fish and Wildlife Service. 2001. *Final Environmental Assessment: Proposed Reintroduction of Migratory Flock of Whooping Cranes in Eastern United States.* Fort Snelling, Minnesota.

United States Fish and Wildlife Service. 2001. *Final Ruling to Establish a Nonessential Experimental Population of Whooping Cranes in the Eastern United States.* Federal Register 66 (123): 33903-33917.

Urbanek, R.P., and D.W. Stahlecker (eds.). 1997. *Proceedings of the Seventh North American Crane Workshop.* North American Crane Working Group, Grand Island, Nebraska.

Walkinshaw, L. 1973. *Cranes of the World.* Winchester Press, New York, New York.

Index

Page numbers in *italics* refer to photographs. Page numbers in **bold** refer to species profiles. Names enclosed in quotation marks (e.g., "Pete") refer to whooping cranes.

Photo Credits

Front cover: National Geographic/Getty Images; front flap: Jennette van Dyk/Getty Images; back cover: International Crane Foundation, Baraboo, Wisconsin (top right, bottom left, bottom center), Mr. Crane Wu (bottom right); spine: International Crane Foundation, Baraboo, Wisconsin

p. 2	International Crane Foundation, Baraboo, Wisconsin
p. 6	Mr. Crane Wu
p. 8	Markus Botzek/zefa/CORBIS
p. 10	L.H. Walkinshaw
p. 11	International Crane Foundation, Baraboo, Wisconsin
p. 12	Brooklyn Museum/CORBIS
p. 16	International Crane Foundation, Baraboo, Wisconsin
p. 20	Martin Withers/FLPA/Getty Images
p. 22	Dorling Kindersley
p. 26	www.operationmigration.org
p. 27	www.operationmigration.org
p. 29	Lynda Richardson/CORBIS
p. 31	Tim Graham/Getty Images
p. 34	National Geographic/Getty Images
p. 37	International Crane Foundation, Baraboo, Wisconsin
p. 40	K.S. Gopi Sundar
p. 45	Arthur Morris/CORBIS
p. 46	International Crane Foundation, Baraboo, Wisconsin
p. 50	Jennette van Dyk/Getty Images
p. 54	Warren Jacobi/CORBIS
p. 58	National Geographic/Getty Images
p. 62	2006 Getty Images
p. 68	Newell Convers Wyeth/Getty Images
p. 71	Courtesy of the National Audubon Society
p. 72	John James Audubon/Getty Images
p. 77	Courtesy of Rich Paul
p. 80	Courtesy of Rich Paul
p. 83	Steve Kaufman/CORBIS
p. 88	Dan Guravich/CORBIS
p. 92	International Crane Foundation, Baraboo, Wisconsin
p. 95	James P. Blair/Getty Images
p. 99	International Crane Foundation, Baraboo, Wisconsin
p. 100	Arthur Morris/Getty Images
p. 105	Courtesy of Rich Paul
p. 110	Dan Guravich/CORBIS
p. 115	Robb Kendrick/Getty Images
p. 122	Time & Life Pictures/Getty Images
p. 123	Time & Life Pictures/Getty Images
p. 126	Raymond Gehman/CORBIS
p. 139	www.operationmigration.org
p. 140	International Crane Foundation, Baraboo, Wisconsin
p. 143	International Crane Foundation, Baraboo, Wisconsin
p. 144	National Geographic/Getty Images
p. 148	Getty Images
p. 154	National Geographic/Getty Images
p. 157	International Crane Foundation, Baraboo, Wisconsin
p. 160	National Geographic/Getty Images
p. 166	www.operationmigration.org
p. 169	www.operationmigration.org
p. 171	www.operationmigration.org
p. 174	Arthur Morris/CORBIS
p. 176	www.operationmigration.org
p. 179	www.operationmigration.org
p. 183	www.operationmigration.org
p. 184	www.operationmigration.org
p. 187	www.operationmigration.org
p. 188	www.operationmigration.org
p. 191	www.operationmigration.org
p. 192	www.operationmigration.org
p. 197	www.operationmigration.org
p. 199	www.operationmigration.org
p. 200	www.operationmigration.org
p. 202	Joel Sartore/Getty Images
p. 207	www.operationmigration.org
p. 209	International Crane Foundation, Baraboo, Wisconsin
p. 211	www.operationmigration.org
p. 212	www.operationmigration.org
p. 214	National Geographic/Getty Images
p. 216	National Geographic/Getty Images
p. 219	Warren Jacobi/CORBIS
p. 221	International Crane Foundation, Baraboo, Wisconsin
p. 223	International Crane Foundation, Baraboo, Wisconsin
p. 225	Martin Harvey/CORBIS
p. 227	International Crane Foundation, Baraboo, Wisconsin
p. 229	International Crane Foundation, Baraboo, Wisconsin
p. 231	International Crane Foundation, Baraboo, Wisconsin
p. 233	International Crane Foundation, Baraboo, Wisconsin
p. 235	Herbert Kehrer/zefa/CORBIS
p. 237	International Crane Foundation, Baraboo, Wisconsin
p. 239	International Crane Foundation, Baraboo, Wisconsin
p. 241	International Crane Foundation, Baraboo, Wisconsin
p. 243	International Crane Foundation, Baraboo, Wisconsin
p. 245	International Crane Foundation, Baraboo, Wisconsin